EDMUNDS, George F. The mayflies of North and Central America, by George F. Edmunds Jr., Steven L. Jensen and Lewis Berner. Minnesota, 1976. 330p ill bibl index 75-39446. 28.50 ISBN 0-8166-0759-1

Will become a classic reference on North and Central America Ephemeroptera. All but three of the twenty families in the world are represented in the Americas. The book provides an updated, well-illustrated key to the adults and nymphs. Once basic morphology is learned, the keys can be used by the non-biologist. A discussion on basic mayfly structure and its biological role precedes the main text. Data on habitats, behavior, and life history are given for each genus. Recommended for college libraries. The serious fresh-water fisherman, entomologists, and limnologists will also find this volume well worth the purchase price.

The Mayflies of North and Central America

George F. Edmunds, Jr., is a professor of biology at the University of Utah, Steven L. Jensen teaches in the department of life sciences at Southwest Missouri State University, and Lewis Berner is a member of the Institute of Biological Sciences at the University of Florida.

The contribution of the McKnight Foundation
to the general program of the University of Minnesota Press,
of which the publication of this book is a part,
is gratefully acknowledged.

The
Mayflies
of
North and Central
America

George F. Edmunds, Jr.

Steven L. Jensen

Lewis Berner

University of Minnesota Press, Minneapolis

Library of Congress Catalog Card Number 75-39446

ISBN 0-8166-0759-1

Acknowledgment of permission to reprint illustrations: Entomological Society of America (Edmunds, Berner, and Traver, 1958; Edmunds and
Traver, 1959; Allen and Edmunds, 1962, 1963, 1965); Entomological
Society of Canada (Allen and Edmunds, 1959, 1963; Allen, 1967, 1974);
Entomological Society of Washington (Jensen, 1974); *Florida State.
Museum Bulletin*, University of Florida (Pescador and Peters, 1974);
Georgia Entomological Society (McCafferty, 1972); Illinois Natural History Survey (Burks, 1953); Iowa State University, Agricultural and Home
Economics Experiment Station, Research Bulletin Division (Fremling,
1960); Kansas Entomological Society (Allen and Edmunds, 1961, 1962;
Edmunds and Allen, 1964; Allen, 1966; Allen and Roback, 1969); New
York Entomological Society (McCafferty, 1971); Pacific Coast Entomological Society (Edmunds, 1960; Edmunds and Koss, 1972); United
States Environmental Protection Agency (Fremling, 1970); University of
Florida (Berner, 1950); University of Utah (Edmunds, Allen, and Peters,
1963). Unpublished illustrations were supplied by R. K. Allen, C. R.
Fremling, W. P. McCafferty, J. G. Peters and W. L. Peters, and A. W.
Provonsha. The balance of the illustrations were prepared expressly for
publication in this volume.

To the late Jay R. Traver,
scholar, teacher, friend.

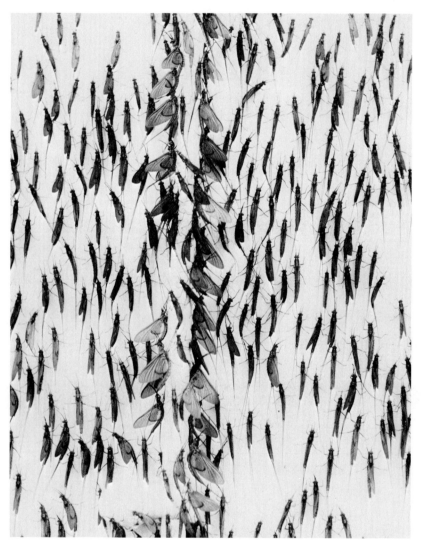

Subimagoes of *Hexagenia bilineata*. (From Fremling, 1960.)

Contents

viii   Contents

Preface

This book was initially conceived by Edmunds and Jensen as an attempt to provide a modern, useful, and well-illustrated key to the adults and nymphs of the genera of mayflies of North America with a brief section on methods. To extend its usefulness we decided to include Mexico and Central America in the geographical coverage. Because the mayfly literature on Mexico and Central America is poorly known, we believed we needed at least a list of species and preferably other data. Even for North America north of Mexico the only recent synopsis of information was Berner's "A Tabular Summary of the Biology of North American Mayfly Nymphs," which was out of print. In the end we decided to pool our resources.

Using the tabular summary as a starting point, we selected data on habitats, behavior, and life history for each genus. We also assembled characters of nymphs and adults for families, subfamilies, and genera and developed brief accounts of extralimital families. The keys to families and the family descriptions are sometimes limited to characters of regional application. The reasons for this are purely pragmatic. Some families resemble others very closely in certain geographic regions; for instance, nymphs of the *Metamonius* group (Siphlonuridae) of Chile, Australia, and New Zealand closely resemble the Baetidae, and the gills of some genera of southern hemisphere Leptophlebiidae are superficially similar to those of genera in other families. We decided not to impose these complex taxonomic separation problems on persons interested primarily in the North and Central American mayfly fauna. We are acutely aware of the need for a more detailed synthetic treatment of the data on nymphal habitats and habits, life history, and adult swarming behavior. We have provided some synthesis, but an adequate synthesis was beyond the scope we envisioned for this work. We have, perhaps unfortunately, mixed unpublished data on biology and distribution with published data. In some cases this is only marginally noticeable, but in other cases the new data exceed all previous information.

ix

In many ways the preparation of this book has been a community effort. We are especially pleased to thank the many persons who have contributed or loaned specimens, allowed us to use drawings or photographs, provided data, given us special aid with certain families, made drawings, or assisted us with the manuscript.

R. K. Allen provided us with data on some Tricorythidae, R. W. Koss on certain Baetidae, W. P. McCafferty on the Ephemeridae, and W. L. Peters and J. G. Peters on the Leptophlebiidae and Behningiidae. Valuable data on *Siphlonisca* were provided by C. P. Alexander. Specimens of considerable use to us were supplied by R. K. Allen, P. H. Carlson, the late W. C. Day, E.-J. Fittkau, J. Jones, R. W. Koss, D. M. Lehmkuhl, P. C. Lewis, G. Longley, W. P. McCafferty, M. Pescador, L. T. Nielsen, J. S. Packer, M. S. Peters, W. L. Peters, A. W. Provonsha, J. R. Richardson, the late J. R. Traver, and P. Tsui. Specimens were also furnished by the following institutions: the Academy of Natural Sciences in Philadelphia, the California Academy of Sciences, the Canadian National Collection, Florida Agricultural and Mechanical University, the Field Museum, and the University of Minnesota.

Most of the illustrations that are not otherwise acknowledged were done at the University of Utah by S. L. Jensen or A. W. Provonsha. Others were done by D. I. Rasmussen, R. K. Allen, Anna H. Kennedy, Nanci Hanson, or G. F. Edmunds at Utah. The cartoons were done by A. W. Provonsha. Anna Kennedy and Dianne Noice provided valuable aid with the manuscript, and Martha Fritter and Natalie Witte helped assemble the plates. The completed manuscript was read by Dr. and Mrs. Peters, P. H. Carlson, P. Tsui, J. Jones, M. Pescador, and M. D. Hubbard, all at Florida Agricultural and Mechanical University.

Many of the figures and much of the data used in this book are spin-offs from studies by Edmunds that have been supported by the National Science Foundation. Other figures and considerable data are from unpublished theses written by S. J. Jensen, R. K. Allen, W. P. McCafferty, and J. S. Packer at the University of Utah under the direction of G. F. Edmunds. The Public Health Service of the National Institutes of Health provided substantial support to L. Berner for his studies of the mayflies of the southeastern United States, and the University of Minnesota made research time available to him for work on the manuscript during two summers while he was teaching in the university's Field Biology Program at Itasca State Park.

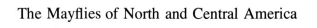

The Mayflies of North and Central America

The Biological Role of Mayflies

The mayflies have fascinated man for centuries because of the brief span of their adult lives. These aquatic insects spend most of their lives as nymphs in water, then develop into winged stages and soon die. The white mayfly, *Ephoron album*, has a winged stage lasting a bare ninety minutes. In fact, the female of this species lays her eggs and dies as a subimago, that is, before ever achieving full adulthood. Similar brevity is almost certainly the case with several other species. Although some species survive as adults for ten days and may live for weeks, most mayflies have an adult life of only two or three days. This brevity is implied in the very name of the order, Ephemeroptera, for the winged stages are indeed ephemeral.

The brevity of the adult stage is possible because the sole function of the adult is to reproduce. The males form swarms which vary with the species from small companies of dancing individuals to spectacular clouds of flying insects. The females enter the swarms, and mating almost always takes place in flight. Within a matter of minutes or at most a few hours the females deposit the eggs in the water. The adults do not even feed. The process of natural selection has resulted in insects whose every adaptation is directed toward the process of reproduction. The mouthparts are vestigial remnants, and in some species the legs are vestigial, except for the forelegs of the male, which are adapted to hold the female during mating in flight; in one genus even the forelegs of the male are vestigial.

The mayflies are almost worldwide in distribution, being absent only from Antarctica, the extreme Arctic, and many small oceanic islands. All but three of the twenty families in the world occur in North or Central America. One of these families, Siphlaenigmatidae, is recognized for a single species from New Zealand, *Siphlaenigma janae* Penniket. The family Prosopistomatidae has a single genus, *Prosopistoma*, which is rather widely distributed in the Old World. The family Palingeniidae contains several genera and is widely distributed in large rivers of the Old World.

Fig. 1. Subimagoes of *Hexagenia bilineata* weighing
down a tree branch. (From Fremling, 1960.)

The nymphs of an Oriental mayfly, *Povilla*, sometimes cause the collapse
of wooden pilings by burrowing into the wood, but mayflies are seldom pests
in the ordinary sense. Great masses of emerging adults (fig. 1) are sometimes
a problem, however, in that the bodies of the insects may accumulate to such a
depth on bridges (fig. 2) that traffic must be stopped until the insects can be
removed by snowplows. But, like snow, mayflies should never be abolished.

The role of aquatic insects as indicators of pollution has received increasing attention in the last two decades. No doubt there have been spectacular mayfly emergences for centuries, but it is almost certain that the great masses of mayflies mentioned in the preceding paragraph are a symptom of man's unknowing interference with the environment. Modern man has enriched streams and lakes with sewage from cities, manure and fertilizers from farms, and natural nutrients from eroding soils. While such overenrichment makes conditions unsuited for most mayfly species, a few forms that strain their food from the water, such as *Hexagenia* (fig. 3), *Ephoron*, and certain caddisflies, thrive on it and greatly increase in numbers. When any river is largely dominated by these filter-feeding insects, it may be on the brink of ecological disaster.

Besides indicating the presence of pollutants, mayflies also help to remove these substances from the waters and to return them to the terrestrial environment. The self-cleansing of a river consists of degrading and removing the organic material. Carbon escapes through respiration, but phosphates and

Fig. 2. *Hexagenia bilineata* attracted to automobile headlights, Mississippi River, Winona, Minnesota. (From Fremling, 1970.)

Fig. 3. Nymph of *Hexagenia bilineata* in artificial burrow. (From Fremling, 1970.)

nitrates are taken out mainly by being incorporated in the bodies of emerging aquatic insects. Calculation of the tons of phosphorus and nitrogen removed from polluted lakes and rivers and returned to the terrestrial segment of the ecosystem by aquatic insects would be of great interest, although at present we cannot make a realistic guess about the amount.

In the 1940s and 1950s Lake Erie was tremendously enriched by a great influx of nutrients, and the number of mayflies which accumulated at the shoreline was incredible. As long as the mayflies harvested much of the influx, there was hope for the natural system. But the capacity of the insects to remove the nutrients was overwhelmed, and great die-offs of mayflies took place during calm periods when the amount of atmospheric oxygen driven into the lake was minimal (Britt, 1955). When this happened, Lake Erie had biologically fallen off a cliff. Britt et al. (1973) report that in the island area of the lake before 1953 averages of about 350 to 400 nymphs per square meter were found with one sample yielding 9,000 nymphs per square meter. Since then there has been a drastic decline to 37 nymphs per square meter in 1957, 46 in 1958, 25 in 1959, with complete elimination by 1962. Wood (1973) estimates the loss of insects in standing crop in the western end of Lake Erie to be 70 million pounds (wet weight). Now the nutrient inflow must be reduced to the point where the lake can recover without the help of *Hexagenia* and

Ephoron. The harvest system of *Hexagenia* and other insects still effectively removes much pollution from the Mississippi River, but only man can determine whether he will allow these insects to continue their beneficial role.

While pollution at least temporarily increases the numbers of certain mayfly species, it causes others in the same area to die out. Thus rivers and streams that once held great biological interest because of their diverse insect faunas have in some cases declined sharply in the last decade. A fuller understanding of the various causes of reduced species diversity and the role of species diversity in aquatic systems is urgently needed. The mayflies constitute a major component of these systems.

Sport fishermen have always been the most interested amateur students of mayflies. The nymphs of the larger species may be used as bait. In the Great Lakes area, for instance, *Hexagenia* nymphs are sold to fishermen. Some skilled fly-fishermen have an enviable knowledge of the emergence times of each species in the area where they fish. Articles now appearing in popular sport magazines and books make it obvious that the devoted fly-fisherman is interested in finding out the scientific names of his insects and in learning their habits in order to obtain catches of fish only dreamed of by others. In the British Isles fly-fishermen have contributed much to the knowledge of the British mayfly species. This British tradition has also been carried to such distant places as the trout streams of South Africa and Tasmania. North Americans have more difficulty relating aquatic entomology to fly-

fishing in this way because, while the British Isles have only forty-eight species of mayflies, North America has more than six hundred species north of Mexico and nearly one hundred additional species in Mexico and Central America. However, many of these do not occur in streams of interest to fly-fishermen, and the common species of aquatic insects in fly-fishing waters in North America can be known on a regional basis.

There is a better reason than fishing for learning about mayflies. Man sees and enjoys nature through his brain more than through his eyes. To walk through a woods or to look into a river without knowing the life that lurks there is comparable to walking through a crowd of strangers. Enjoyment comes with recognition of what we see. One of our friends, a man who had spent many years in an area and knew every road and stream, had not realized that "all those fascinating little bugs" were under the rocks in the streams. His world was much richer after only a week of working with us in the field.

METHODS

Collecting Nymphs

Securing a diversity of mayfly nymphs requires persistent work in a wide variety of habitats using many different collecting devices. Because Ephemeroptera nymphs are microhabitat specialists, any one species may occur only in an extremely small fraction of a given habitat. Some mayfly nymphs may be so elusive that several man-days are required to locate the species in a short stretch of stream, while in large rivers and lakes certain species may escape detection for years. Thus, the more diverse the habitats sampled and the methods used in a particular body of water, the more species will be collected. The most effective collecting techniques devised by various mayfly specialists of the world will be described in this section.

Collecting Devices and Procedures

Hand screens (fig. 4). The most versatile collecting tool is the hand screen. It consists of a piece of screen bolted between two handles. Perhaps the most useful size of screen is 18 inches high by 24 inches wide, bolted between two 36-inch wooden dowels of ¾-inch diameter, the dowels being split for 20 inches of their length. For collecting in deeper water, screens 24 to 30 inches high and of varying widths are required. Screen material can be steel, bronze, aluminum, nylon, or fiber glass. Of these, bronze seems to be the most durable and the easiest to handle. Nylon is available in the greatest variety of mesh sizes. Because cut screen frays, it is essential that the bottom edge —

and the top, if possible — have selvage or woven edges. The cut edges at the sides are folded double and bolted into the slots of the handles.

Fig. 4. Hand screen in use.

Apron nets (fig. 5). The apron net consists of a triangular scoop on the end of a long handle. The scoop has coarse mesh on the top and fine screen in the bottom. The coarse mesh keeps heavy vegetation and debris from entering the scoop while the fine mesh retains the specimens in the bottom. The specimens are taken out for examination through a trap door at the back of the scoop.

Fig. 5. Apron net.

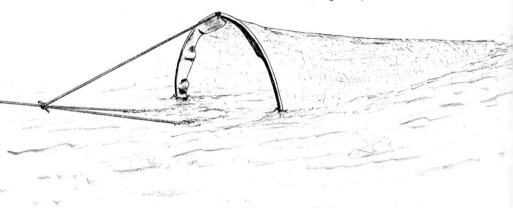

Fig. 6. Small drift net.

Drift nets (figs. 6, 7). Drift nets are underused but valuable devices for collecting drifting insects in streams. The nets vary in overall size, complexity, and mesh size. Heavy drift nets in the current of deep rivers may yield rare species. An inexpensive nylon net with a supporting wire or strip of bamboo in the mouth is often effective (fig. 6). The drift net can be used to take adult specimens along with nymphs if floats are placed at the waterline. Such a net is easily attached with a triangular bridle to a rock or an overhanging limb. Drift sampling can also be done with a hand screen or a regular insect net. Larger drift nets (fig. 7) can double as seines.

Dip nets and scrapers. Dip nets of various designs are useful for sampling certain aquatic habitats. They are not as heavy or as cumbersome as apron nets, and they can be used in shallow areas. A dip net usually consists of a

Fig. 7. Large drift net.

Fig. 8. Drag screen in use on soft bottom.

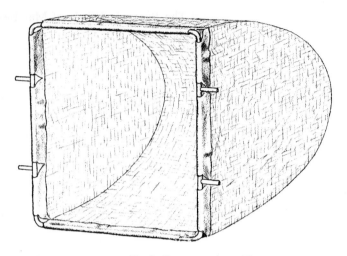

Fig. 9. Drag screen in position.

simple net bag of fine mesh with a wire frame and a long handle. Vladimir Landa (Prague, Czechoslovakia) uses a metal cup with a screen bottom and a long handle.

Scrapers are much heavier than dip nets, usually consisting of a square or rectangular metal frame on a long handle, with net or screen of varying mesh size lining all but the upper opening. Rakelike teeth on the frame disturb the stream bottom while the scraper is being used.

Edmunds designed a drag screen (figs. 8, 9) for use in large rivers after he discovered that many active nymphs were escaping over the tops of hand screens. A net opening of 36 by 36 inches is effective for a drag screen, but other sizes can be built according to one's needs. Two persons grasp the handles at the sides and move the screen upstream rapidly, scraping it along · the bottom to disturb nymphs and wash them into the net bag. On soft, sandy, or muddy bottoms the net can be turned so that one set of handles penetrates into the substrate while the bottom is disturbed upstream (fig. 9).

The common 6-inch-diameter household strainer is very useful for collecting nymphs. It can be used in places that cannot be reached with larger nets, and it is highly versatile in its applications. The insects may be picked directly from the strainer, or the material in the bottom may be dumped into a white pan for sorting.

Dredges. A number of commercially made dredges of heavy construction, such as the Ekman and Peterson dredges, can be used in deep waters. The dredge, with its heavy traplike jaws open, is lowered to the bottom on a nylon rope. A metal messenger is dropped which triggers the spring-activated jaws to slam shut on a sample from the top layer of substrate.

Miscellaneous devices. Aquatic light traps of various designs may be useful for collecting phototropic mayfly nymphs, although these traps seem to be more effective with other aquatic insects. Drag-type bottom samplers are similar to scrapers in function except that the weighted frame and net are dragged across the bottom on a long rope. Brush traps made with bundles of trash wood and branches inside a wire screen cage are especially effective where inhabitable substrates are uncommon. The cage is submerged on the end of a recovery cable attached on the shore or to an anchor and left in place until it is colonized by insects. Other artificial substrates such as laminated plates, rock-filled baskets, webbing mats, or artificial U-shaped burrows which nymphs can colonize are also used for either quantitative or qualitative sampling.

The electro-seine or fish shocker is a workable collecting device, but the electrodes must be held close together and a screen must be placed downstream to intercept the dislodged nymphs. In most areas such devices require permits from the fishery management officials, but if they are used judiciously, with the electrodes nearly together, most fish escape unharmed. A disadvantage of the electro-seine is that it is quite selective, yielding nymphs of some species in greater numbers, and others less commonly, than does the hand screen in the same habitat.

The riffle net (fig. 10) used by E. F. Riek (Canberra, Australia) consists of a piece of nylon netting stretched with a slight sag over an 18-inch ring with a 6-inch (or longer) handle. This net may be substituted for a hand screen or an apron net in heavy plant growth. It is almost as versatile as the hand screen and is better than the hand screen in some habitats.

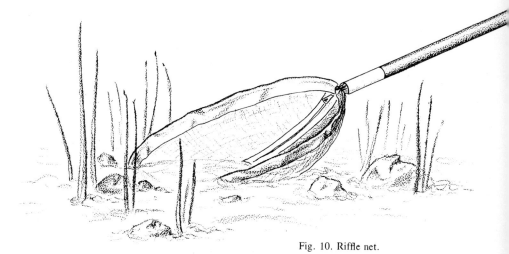

Fig. 10. Riffle net.

Habitat Collecting Methods

Streams and rivers. The greatest species diversity occurs in these habitats, and numerous collecting methods are effective. The hand screen is generally used in running water. The screen is held across the current at the bottom of the stream to catch nymphs dislodged when the substrate upstream is disturbed. If two people are collecting, one holds the screen while the other stirs the bottom upstream; if the collector is working alone, he must hold the screen while he disturbs the bottom with his feet or with a small rake. In pools or slow-moving water at river edges the collector moves the screen directly and rapidly toward shore.

In habitats where a hand screen cannot be maneuvered, a dip net is often profitable. The rootlets and debris near undercut banks frequently harbor certain mayflies, and vigorous action is required to dislodge them.

The use of drift nets in stream and river habitats often yields species not readily taken otherwise. These nets also accumulate the cast skins of molting nymphs and subimagoes which have partially emerged from the nymphal skin. This often allows association of the nymph and the adult. Drift nets should be used for relatively short periods or they will be clogged by the drifting material; if secured properly, they can be used overnight. Long series of nymphs of many genera, including the burrowers and the rare *Pseudiron*, have been obtained from overnight accumulations in drift nets. Daytime collecting with these nets is most productive during nymphal emergence periods. The nets are also valuable in mountain areas where species float down from hard-to-reach upstream habitats.

Many mayfly genera are restricted to large rivers, which are the source of most of the exciting and unusual genera taken in North America in the last twenty years, and which are also undoubtedly the most difficult habitats to sample. The depth and the strong currents often preclude sampling except near the edge where the water is shallow. In some cases the only way to sample the deeper areas is to dredge them from a boat, but drift nets will often collect many species in these areas. Unfortunately, the microhabitat data cannot be determined for species collected in this manner. A brush trap or other submerged device will be colonized by some kinds of mayflies. When such traps are carefully removed and washed off after several days, they frequently yield nymphs of species that are otherwise difficult to obtain from large rivers.

Ponds and lake margins. Collecting in these habitats involves slightly different techniques since there is no current to carry specimens to the collecting device. One very effective device for sampling such habitats, especially in areas with abundant aquatic vegetation, is the apron net, which may be maneuvered through the vegetation or scraped along the bottom to dislodge nymphs. Heavy-mesh screen on the top permits specimens to fall into the net but keeps heavy vegetation and debris out.

Dip nets, which are readily maneuverable in shallow water, and riffle nets are also very effective in these habitats. Hand screens, although generally more cumbersome, can be moved through the vegetation in the same way one uses an apron net or a dip net, or they can be passed through the debris stirred up from the bottom.

Collecting Adults

A wide variety of collecting techniques is necessary to get a good representation of adult mayfly specimens — the more diverse and imaginative the collecting methods, the greater the collector's success.

Tent traps such as the Malaise or Townes trap often yield large numbers of adult mayflies. Curtains of netting placed over a stream serve as a surface on which the mayflies alight.

The capture of flying mayflies, particularly swarming males, requires a light, maneuverable net with a very long handle (preferably one that can be

Help!

adjusted in length). The net should have a large throat and a fairly stiff bag. The BioQuip tropics net (fig. 11) is excellent, especially if the handle section is covered with plastic tape. An equally good handle can be made by cutting bamboo between the nodes and taping the cut ends. By swinging the long-handled net at random over a stream after dark, the collector often can secure examples of species which cannot be taken earlier in the evening. Swarming males that stay tantalizingly out of reach at other times are often driven down toward the water by gusts of wind. Some species swarm too high, however, for any known netting procedure.

The search for swarming males should not be limited to stream and lake margins. Males of certain species swarm around meadows, bushes, trees, and forest openings, sometimes a mile or more from water. Others swarm over areas that are lighter or darker than the general background. Mayflies often swarm over highways near water. Among the species that swarm over water, some are attracted only to white water or to slicks on the surface.

Collecting mayflies at lights is frequently productive, except in areas where the temperature drops quickly in the evening. In these areas collecting at lights is always better under overcast skies because in such circumstances the temperature drops more slowly and low light intensities occur earlier. For some species collecting at lights is effective only very early in the evening. Various lights are effective, but generally black and mercury vapor lights are best. Several kinds of lights used together yield the greatest number of species, but even under ideal conditions (warm, humid evenings), lights never seem to attract all the species known to be in an area. Paul Carlson, Jerome Jones, and others at Florida Agricultural and Mechanical University confirm that adults and subimagoes of some species appear at lights only at certain times during the night. Therefore, all-night light collecting runs are necessary for a complete sampling of species.

Subimagoes may be netted one at a time as they emerge from the stream. Because many mayflies are taken as subimagoes, and because the taxonomy is based largely on imagoes or nymphs, the subimagoes must be allowed to molt to the imago stage. The subimago stage lasts from a few minutes to a few days in the various species, but for most the stage lasts from eight to twenty-four hours.

Netted subimagoes are very fragile and should be gently transferred to subimago boxes immediately. The preferred way to accomplish this is to place a small glass-bottomed box (fig. 12) over the subimago. The insect will usually move to the apparent escape route much more readily than it would enter a closed box or sack, and handling is therefore unnecessary. If it becomes necessary to handle the subimago, the insect should be picked up very lightly by the wings, using flat, pliable forceps such as those sold for handling

Fig. 11. Long-handled net.

butterflies. Subimago boxes are available in four nesting sizes ranging from 1⅛ to 2¼ inches in diameter. Ideally, the boxes should be stored in a cool, humid, insulated container. In any case they must be kept away from heat and sun, for if they are turned with the glass to the sun for only a few seconds, the subimagoes may die.

Rearing

Most of the life of a mayfly is spent in the nymphal stage, and this stage is most frequently represented in collections. Nevertheless, most taxonomic studies have been based on adults. This situation is slowly being corrected, but it is still impossible to determine the species of nymphs in most genera. In Mexico and Central America it is not even always possible to determine the genus of nymphs. This is particularly true in the family Baetidae and to a lesser extent in the family Leptophlebiidae. The species are gradually becoming known, however, through the efforts of researchers who rear nymphs to the adult stage.

Two methods of rearing can be used. One is mass rearing, in which many nymphs presumed to be of one species are placed in an enclosure where the adults can be captured and identified. This frequently yields numerous adults and allows for absolute association of the nymphs and the adults. However, in most cases reliable association of nymphs and adults is accomplished by isolating single individuals in cages. Nymphs with black wing pads are nearly mature, and by selecting these, only a day or two is required for rearing. This requires relatively large numbers of simple, inexpensive cages. Various cages have been devised, but perhaps the simplest and most effective one is a slight modification of a cage first constructed by Müller-Liebenau (1970) out of a plastic cup with a lid (fig. 13). Two openings are cut in the lower half of the cup, and a large opening is made in the lid. The entire inner wall of the cup is then lined with nylon mesh or fine-mesh fiber glass screen, and another piece of screen is cut to fit the lid opening. The container is then placed in a rearing tank, a wire basket, or a float (fig. 14) in the stream. It is usually necessary to shade the cages from the sun. A few tree branches placed over the cages will provide adequate shade and also will disguise the cages so they are less likely to be disturbed by curious passersby. The screened top opening prevents the buildup of heat and allows the collector to scan the cages quickly to see if the subimagoes have emerged. We recommend lining the entire side wall of the cup with screening to allow the subimago to cling to the side; without the screening the insect may slip back into the water and drown. When the subimago emerges, it must be carefully placed in a box to transform, or the original rearing cage may be used as the subimago cage. The wings of the subimago should be touched only very gently and with very soft forceps.

Fig. 12. Subimago box cage.

Fig. 13. Rearing cage (plastic cup).

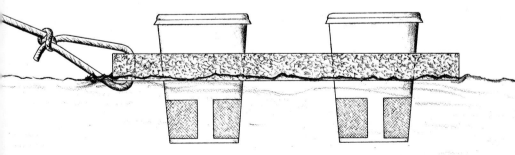

Fig. 14. Rearing cages in float.

There are advantages and disadvantages in both field and laboratory rearings. To rear nymphs in the laboratory, a suitable water supply must be available. Tap water is usually not of adequate quality for rearing unless it is first dechlorinated. This necessitates recirculating the dechlorinated water through filter systems. Laboratory rearings are particularly effective for species that can withstand slower currents and higher temperatures. If the rearing tanks are refrigerated, a greater variety of species can be reared. Fish hatcheries are ideal for rearing because the cages are protected from intruders and local personnel may cooperate by lifting out cages containing subimagoes to prevent accidental drownings. Rearing is also feasible at water treatment plants and a variety of other facilities. Whenever nymphs are transferred to other waters for rearing, emergence may be delayed or prevented if the specimens are placed in cooler waters. Perhaps the greatest drawback to field rearing, especially in areas distant from the laboratory, is that the worker must stay in the area, often for an extended period of time.

When nymphs are collected in the field, they must be very carefully transported to the laboratory. Some species may simply be carried in a bucket of water, but others are transported better on wet burlap or in damp sphagnum moss. Portable aerators powered by batteries are usually necessary in the transportation of mayflies in water unless the water is agitated sufficiently by the vibration of the vehicle in which the container is carried.

For rearing and various other purposes, it is frequently desirable to know the sex of mayfly nymphs, and the sex of half-grown nymphs of most species can be determined. The eyes of the male are usually larger than those of the female and often are different in shape. Furthermore, the developing male genitalia gradually become visible as outgrowths of the ninth sternum. In Caenidae, Tricorythidae, and some other mayflies there is little or no sexual dimorphism of the eyes. In many of these the cerci (and also the terminal filament in most species) are often much longer in the male imago than in the female. In the nymphs this is frequently expressed by the appropriate caudal filaments being much thicker (especially at the base) and more strongly tapered in the male than in the female.

Preservation of Specimens

Adults. The preservation of adult mayflies is a controversial subject; some workers prefer to preserve the adults in alcohol and others prefer to pin them. We think that when a series of specimens is available, a few should be pinned and others should be placed in alcohol. The advantages of pinned specimens is that their color is preserved better and they last longer, judging from experiences with museum specimens. We have examined specimens pinned by B. D. Walsh in the 1860s which are still in good condition. However, pinned specimens are exceedingly fragile; because of the likelihood of breakage, it is best to store them in small unit trays with polyethylene foam bottoms. The specimens are easily removed from the trays, and any broken parts are usually found directly below the specimens from which they came. Making slides of certain small mayfly parts, such as the male genitalia, is much harder with pinned specimens than with those preserved in fluid.

For the preservation of adults in alcohol we recommend 80% ethyl alcohol with 1% Ionol (an antioxidant) added. The Ionol lessens the bleaching of the specimens with age and keeps them wettable in case the alcohol should accidentally be allowed to evaporate from the container. A. W. Provonsha (personal communication) states that adults and nymphs keep their color if freshly collected specimens are subjected to heating, almost to the boiling point, for a few minutes. Nevertheless, adult mayflies preserved in alcohol are very fragile, and the procedures described for preserving nymphs are also recommended to prevent damage to adult specimens.

Nymphs. The most common preservative for field collecting of nymphal mayflies is 95% ethyl alcohol. Strong alcohol is used rather than a more dilute form because when nymphs are collected into the vial considerable water is usually introduced with them. Although alcohol as a preservative is reasonably acceptable, there are better fluids. We prefer to collect specimens directly into modified Carnoy fluid (glacial acetic acid, 10%; 95% ethanol, 60%; chloroform, 30%), draining off the Carnoy fluid within twenty-four hours, when feasible, and replacing it with 80% alcohol. When this is not possible, no particular damage is done by leaving the specimens in Carnoy fluid longer. An excellent substitute for Carnoy fluid is Kahle's fluid (formalin, 11%; 95% ethanol, 28%; glacial acetic acid, 2%; water, 59% — to be added in the field). This preservative is particularly advantageous for field trips when the weight of the preservative is important because 59% of the fluid, the water, need not be carried. Kahle's fluid also should be drained off the specimens within a week and replaced with 80% alcohol. With both Carnoy fluid and Kahle's fluid, the nymphs keep their color well and are firm, and their legs, antennae, and gills are not detached as readily. Such specimens can be used for studies of internal anatomy.

Because both nymphs and adults are fragile, jostling of the specimens may result in the loss of legs, gills, antennae, or caudal filaments. During field transportation the vials should be completely filled with preservative, leaving no air bubbles. Our personal preference is to collect the specimens into vials which are placed in widemouthed glass or plastic jars only slightly taller than the vials (figs. 15, 16). As each vial becomes half filled with specimens, an identifying label is placed inside. The vial is then dropped into the jar of fluid, and the neck of the vial is plugged with a piece of thin plastic film such as

Saran, followed by a wad of cotton or a piece of cellulose sponge. A teasing needle or forceps should be used to force out the last air bubbles. In this way even violent movements of the jar will result in little or no shaking of the specimens in the fluid. A bubble of air in a vial of fluid, however, is as destructive to the specimens as a rock of the same size would be. During shipment the jars of vials or the separate vials should be packed carefully in a sturdy container.

Labeling of Specimens

Although labeling mayflies is no different from labeling other insects, a few widely used techniques are worth recording. For specimens in alcohol, all labels are placed around the inside circumference of the vial with no overlap so they may be read with the vial in a normal upright position (see fig. 16). A microelite typewriter (which produces seventeen characters per inch) is used

Fig. 15. Field preservation.

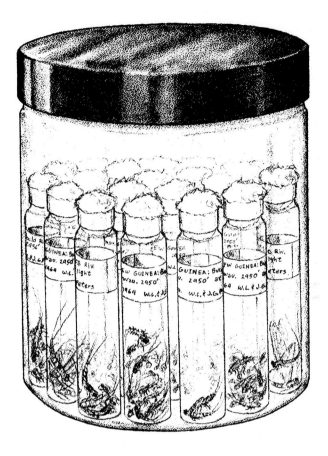

Fig. 16. Labeled material for study or storage.

to prepare the labels on offset master mats, and a large supply is printed for each locality. The labels are printed on rag bond paper (preferably twenty-pound stock).

With the ink used in our duplicating machines, we have found it necessary to dry the labels at 120° F for one month before immersing them in alcohol, but some duplicator inks will dry in twenty-four hours. The drying time and the permanence of the duplicated labels should be tested. We have tested labels duplicated by xerography and have found that there is no apparent loss of clarity in over five years. Duplicated labels or typewritten labels can be used immediately if they are washed for a few minutes in 5% acetic acid and alcohol to remove excess ink or if they are dried at high temperatures with good ventilation.

Good clean copies of the labels are photographically reduced and made into metal master mats for printing microlabels. The microlabels are used to label pinned specimens and microscope slides of specimens. They can be attached to standard slide labels with a white glue that becomes clear as it dries.

Entomologists frequently prepare brief labels in the field and write more thorough labels when they return to the laboratory. This practice has inherent dangers which must be weighed against the advantages of rapid labeling in the field. Those who use code labels run the risk of losing data, but a few precautions will minimize this danger. Before each field trip labels can be prepared giving the area, the month, and the year; the labels for various lots of specimens are then given consecutive penciled code numbers to ensure that series of code-labeled materials from different field trips are not mixed. In the field the full label for each locality number is written in a notebook with a carbon copy, and the locality and the field number are entered on a map. The carbon copies are mailed back to the laboratory, and the map and field notebook labels are kept separate so that if one set of records is misplaced the data are not lost.

The information to be included on a label once the specimens are in the laboratory depends on the type of study being conducted. However, in all cases it should include at least a precise geographic reference to the collecting site, the collecting date, and the collector's name. Other data such as water temperature, nature of substrate, and current velocity are optional. Figure 17 shows some examples of data tags used by the authors.

UTAH: Uintah Co., Uintah Riv.
at Fort Duchesne
11-IX- 71 L. S. George

UTAH: Green Riv., Split Mtn.,
Dinosaur Nat. Mon.
12-IX-70 G. L. Steven

UTAH: Summit Co., 1 mi. N. W.
Peoa, 6100' elev.
16-IX-71 G. S. Lewis

UTAH: Uintah Co., Uintah Riv.
at Fort Duchesne
11-IX- 71 L. S. George

UTAH: Green Riv., Split Mtn.,
Dinosaur Nat. Mon.
12-IX-70 G. L. Steven

UTAH: Summit Co., 1 mi. N. W.
Peoa, 6100' elev.
16-IX-71 G. S. Lewis

Fig. 17. Labels for specimens in vials and microlabels for pinned specimens or microscope slides.

Preparation of Slides

Mayfly parts must frequently be studied from slides. Most structures can be easily mounted on slides as outlined below. Some parts such as nymphal gills are usually best studied in alcohol, but for most structures a better understanding can be achieved by studying both a slide specimen and an alcohol specimen. The proper instruments must be used in handling and studying alcohol specimens and in the preparation of slides. Fine-pointed forceps of the sort available from jewelers' supply companies are essential. Fine-pointed needles are also necessary; these can be made easily by cementing *minuten* pins into the needle orifice of a fine-gauge (25-gauge or smaller) disposable plastic syringe; commercial jewelers' pin vises may serve equally well as handles. The final honing of points or blades is best done on a piece of leather strapping.

The wings of adult specimens can be mounted by floating them from clean alcohol onto the slide, arranging them properly in a thin film of alcohol, and covering them with a square cover slip (approximately 22 mm). Narrow strips (about ¼ inch by 1½ inches) of white gummed paper are much better than any self-adhesive paper or tape for holding the cover slip tight to the slide. We have never found a tape suitable for this purpose that does not require wetting, and ordinary sticky office tapes are clearly to be avoided. After the wings are in position on the slide, the alcohol is then allowed to evaporate. Crumpled wings from dry specimens can be flattened by dropping them on gently boiling water and then quickly floating them onto a slide. Mayfly wings are best studied as dry mounts; wings mounted in Canada balsam generally show fewer details.

Male genitalia from dried specimens may need to be softened before mounting. A satisfactory mount results when they are placed for one or two hours in a solution of 10% potassium hydroxide or sodium hydroxide and then are dehydrated and mounted in Canada balsam. For male genitalia or other parts of adults or nymphs preserved in alcohol, we prefer to avoid hydroxides except for very large or dark specimens that otherwise would not clear. Structures can be mounted directly from 95% alcohol or cellosolve (ethylene glycol monethyl ether) into specially prepared Canada balsam. Commercial neutral Canada balsam in xylene is allowed to dry until it is highly viscous; it is then returned to suitable consistency by replacing the evaporated xylene with cellosolve. Structures placed in this mixture may cloud temporarily but seldom for more than an hour. The clouding can be reduced by passing the structures through pure cellosolve before mounting.

The cover slip for balsam mounts should be no larger than necessary. For small structures such as male genitalia and claws we use 8-mm round cover slips; for larger structures we use 12-mm to 18-mm round cover slips. With the small cover slips it is easier to find small structures under high magnifica-

Fig. 18. Labeled slide.

tion, specimens are not flattened when the balsam dries, and more than one structure or set of structures may be mounted on a slide. When the structures to be mounted are large or thick, the cover slip can be supported with commercial rings, fine glass rods, or short pieces of monofilament fishing leader of appropriate diameter.

An effective method for positioning mounts is to put the structures in a thin film of balsam on the cover slip or the slide. The parts can be repositioned periodically until the balsam has become quite firm. The cover slip can then be transferred to the slide with additional balsam. The structures must be completely covered with balsam and allowed to dry in a dust-free place or in a petri dish. Regardless of how the balsam mounts are made, they must be stored flat for at least a year.

It is essential that slides be cross-referenced to the specimen from which they came with a microlabel in the vial (or on the pin of a dry specimen) which might read "wings on slide" or "genitalia on slide." The slides should also be fully labeled (fig. 18) with the locality, the date, and the name of the collector or referenced to a specific specimen. When drawings are made for publication, the slide or the specimen used should be so labeled.

SYSTEMATICS

Classification System

The system of classification used in this book is modified from that proposed by Edmunds and Traver (1954) and used by Edmunds, Allen, and Peters (1963). In Needham, Traver, and Hsu (1935), and in many entomology textbooks, the mayflies are divided into only three families. However, most European entomologists have recognized as families the groups that North American mayfly workers have called subfamilies. Our reason for not recognizing the three-family system is based on the fact that the classification of Needham brings together in one family distantly related groups. Needham's

Heptageniidae is monophyletic; the Ephemeridae is diphyletic because of the inclusion of the Neoephemeridae; and the family Baetidae is an assemblage of forms that is undefinable except by default (that is, by their failure to have the characters of the Heptageniidae or the Ephemeridae in the sense of Needham). Edmunds (1962a) has discussed the rationale for recognizing the larger number of families used in this work.

In studies of the relationships of the families the prefixes *proto-* and *pre-* are used regularly to refer to ancestors of extant forms. If we wish to indicate that we believe that the ancestors of family B (if they were at hand as fossils) would have had the characters of family A, we refer to the ancestors as proto-A. As an example, we believe that the ancestors of the Caenidae would have had characters much like the existing genus *Potamanthellus*, a typical member of the Neoephemeridae. If, however, we believe that families A and B arose from a common ancestor that had not yet evolved the characteristics of either A or B but was more similar to A, we would refer to this ancestor as pre-A. An example of this kind is seen in the hypothesized ancestry of the Heptageniidae; the ancestor of this family must not yet have evolved the characteristic filter-feeding adaptations of *Isonychia*, but the lineages leading to *Isonychia* and the more modified Heptageniidae probably arose from the same ancestor.

The present family Siphlonuridae has among its members the largest number of primitive characters, although no one genus illustrates all the primitive characters. The existing Heptagenioidea appear to be families derived from siphlonuridlike ancestors. The relationship of the other superfamilies to the Heptagenioidea is somewhat obscure. The Prosopistomatoidea and Caenoidea are clearly derived from a common ancestor, and some workers believe the two superfamilies should be combined. The origin of the Caenoidea, however, is quite obscure, although they must have been from early proto-Heptagenioidea. The Leptophlebioidea and the Ephemeroidea show a close relationship. The line may have originated from proto-oniscigastrine Siphlonuridae, a small subfamily now extant only in Chile, Argentina, Australia, and New Zealand. The Leptophlebiidae appear to have arisen from a common ancestor that also gave rise to the Ephemerellidae and the Tricorythidae. The proto-Leptophlebiidae gave rise to the Ephemeroidea, and the nymphs of Leptophlebiidae are clearly more similar to the Ephemeroidea than to the Ephemerellidae. The adults of the Leptophlebiidae tend to show more similarity to the Ephemerellidae. A family such as the Leptophlebiidae which shows both pregroup and postgroup relationships tends to pose a classification problem because there are grounds for placing them either with the pregroup relatives (as we have done), in a separate superfamily, or with the Ephemeroidea, a postgroup superfamily.

Most mayfly workers throughout the world are in fairly close agreement on the composition of the families, but there is a growing tendency to split the stem group which we call Siphlonuridae. There is very little agreement on the relationship among the families or the composition of the superfamilies. There are similar problems concerning the recognition of genera, whether or not to recognize certain subgenera, and whether or not to elevate subgenera to generic rank. In addition to the subfamilies that we recognize, others have been proposed. Further, some of the groups that we call subfamilies have been recognized as families. A few subfamilies have been placed in other families, and some genera have been placed in other subfamilies.

Nonspecialists frequently ask "What classification is correct?" The answer is that any of several classifications may be correct (based on the current state of knowledge), but some can be obviously incorrect because they group distantly related forms into a given classification group (genus, subfamily, or family) and exclude more closely related forms. Classification depends not only on factual data such as structure, behavior, and the direction of its evolution, but also on the principles that are used in constructing a classification from this knowledge.

Increases in knowledge of the mayflies lead to many changes in classification, and it is important to realize that such knowledge is not equally available or known to all persons who decide on what classification to use. An example of changes brought about by increased knowledge is seen in the mayflies that we place in the families Tricorythidae and Caenidae. These were long regarded as one family, Caenidae (or as the subfamily Caeninae). Studies of the characteristics of these and other mayflies have shown in a very convincing manner that these two families are distantly related. The obvious similarities of *Tricorythodes* (Tricorythidae) and *Caenis* (Caenidae) resulted from convergent evolution caused by similar natural selection pressures on the two groups. In Africa, some of the Tricorythidae are similar to the Ephemerellidae, their true close relatives. The Caenidae are now known to be most closely related to the family Neoephemeridae. Since 1920 additional knowledge has made it increasingly clear that the Tricorythidae are distantly related to the Caenidae and cannot be correctly grouped with them.

Another common source of classification differences arises from the fact that different workers have different philosophies about grouping forms, and there is no exclusively correct set of principles. For example, mayflies of the genus *Isonychia* (subfamily Isonychiinae) are usually placed in the family Siphlonuridae because the adults have characteristics shared with other Siphlonuridae. In contrast, the nymphs of *Isonychia* share more characters with the family Oligoneuriidae. We believe that *Isonychia* evolved early in geological time from primitive Siphlonuridae and that an *Isonychia*-like mayfly gave rise to the ancestral Oligoneuriidae. *Isonychia* has been classified variously in the Siphlonuridae, in the Oligoneuriidae, and as a group equal to the rest of the North American Siphlonuridae. Obviously *Isonychia* can be placed in a group by itself, but this gives little information about its relationships. *Isonychia* can also logically be placed in the Siphlonuridae because its ancestors were similar to Siphlonurinae or in the Oligoneuriidae because the Oligoneuriidae had ancestors similar to *Isonychia*. Three mayfly workers could agree about the evolutionary relationships of these groups and arrive at three different classifications, none of which is incorrect.

Another cause of variation in classification is the level at which subfamily or family categories should be applied. There are other considerations as well, but the ones we have noted are the main reasons why classification schemes do not remain uniform and stable.

We have not attempted to characterize the superfamilies or to key them because the range of characters is so great within these taxa that they are not particularly useful as identification units. Derived superfamilies such as the Ephemeroidea and Prosopistomatoidea are easily characterized, but those containing primitive groups, as does Heptagenioidea, are difficult to characterize. Edmunds takes responsibility for the opinions expressed concerning the relationships of the families. All three authors are, however, in agreement about the familial and subfamilial classification used in this volume. The hypothesized phylogenetic relationships of the subfamilies are shown in an end-view diagram (fig. 19).

In the systematics section of Needham, Traver, and Hsu (1935) Traver listed 507 species as occurring in North America north of Mexico. Of the species names then considered valid (including several that were stated to be doubtfully valid), some have been placed in synonymy and others have been reduced to subspecies rank. Description of additional species during this time, less synonymies, has now raised the number of named mayfly species in North America north of Mexico to 622. Some of these species extend into Mexico, but an additional 89 named species in Mexico and Central America are indicated in our tables. There are a number of cases where unnamed or unidentified species of genera largely confined either to South America or to the United States and Canada are included in the tables to indicate their occurrence in southern Mexico or Central America. Undoubtedly most of these are undescribed species. Although Mexico and Central America perhaps have a larger percentage of their mayflies unnamed than do other regions, there are undescribed mayfly species throughout the region covered by this manual.

Traver placed the species occurring in North America North of Mexico in 47 genera; for the same region we recognize 60 genera. We treat an additional

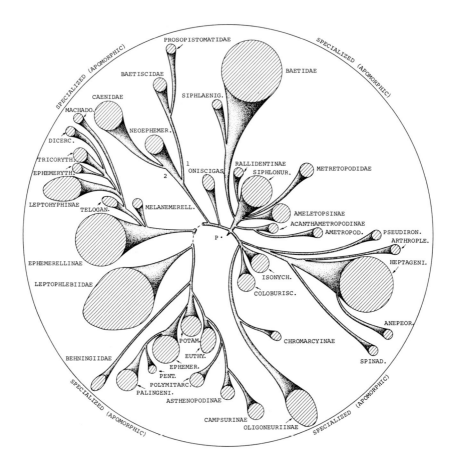

Fig. 19. End-view phylogeny diagram. The center (p) is the point of the primitive ancestor (with plesiomorphic characters). Increasing specialization (with increase in apomorphic characters) is indicated by the distance from the center. The extant diversity of each family or subfamily is indicated by the area of the end of the lineage. Fossil families are excluded because of extreme disagreement on their recognition and position. The chart can be followed in a modified clockwise pattern. The Siphlonuridae are represented by the Oniscigastrinae, Rallidentinae, Siphlonurinae, Ameletopsinae, Acanthametropodinae, Isonychiinae, and Coloburiscinae. Other heptagenioid families are Siphlaenigmatidae, Baetidae, Metretopodidae, Ametropodidae, Heptageniidae (with subfamilies Pseudironinae, Arthropleinae, Heptageniinae, Anepeorinae, and Spinadinae), and Oligoneuriidae (with subfamilies Chromarcyinae and Oligoneuriinae). The Ephemeroidea include Polymitarcyidae (with subfamilies Polymitarcyinae, Asthenopodinae, and Campsurinae), Euthyplociidae, Potamanthidae, Ephemeridae (with subfamilies Ephemerinae and Pentageniinae), Palingeniidae (see footnote with description of this family), and Behningiidae. The Leptophlebioidea include Leptophlebiidae, Ephemerellidae (with Ephemerellinae, Teloganodinae, and Melanemerellinae), and Tricorythidae (with Leptohyphinae, Ephemerythinae, Tricorythinae, Dicercomyzinae, and Machadorythinae). The Caenoidea include Caenidae and Neoephemeridae, and the Prosopistomatoidea include Baetiscidae and Prosopistomatidae. The abbreviations should be self-explanatory.

7 genera that occur only in Mexico and Central America. The genus *Thraulus*, which seemingly makes a total of 68 genera, is included in our text only for two Central American species whose true affinities are obscure; there is no evidence that true *Thraulus* occurs in the New World.

Three of the genera treated by Traver are now synonyms. *Neocloeon* is a synonym of *Cloeon*, *Blasturus* of *Leptophlebia*, and *Oreianthus* of *Neoephemera*. Further, *Iron* and *Ironopsis* are subgenera of the formerly Palearctic *Epeorus*.

The increase in the number of genera for North America north of Mexico has resulted from the following changes: Seven genera have been named from species unknown in 1935 (*Analetris, Edmundsius, Apobaetis, Paracloeodes, Dactylobaetis, Dolania*, and *Spinadis*). The genus *Stenonema* has been split with part of the species now in *Stenacron*. One species has been removed from *Hexagenia* and placed in *Litobrancha*. The North American species of *Thraulus* were placed in *Traverella*. A newly discovered species was placed in a new genus *Metreturus*, but it was subsequently found that in a paper then obscure in North America the genus had been described as *Acanthametropus* from a Siberian species. The generic name *Tortopus* was considered a synonym of *Campsurus* by Traver, but it is now recognized as valid.

Five genera — *Baetodes, Homoeoneuria, Lachlania, Leptohyphes*, and *Homothraulus* — which were unknown north of Mexico in 1935 are now known from the United States, and *Lachlania* is known also from Canada.

Using the classification proposed by Needham (in Needham, Traver, and Hsu, 1935), Traver arranged the genera in 14 subfamilies and 3 families. We recognize 24 subfamilies and 17 families. In a general way the families recognized herein are equivalent to the subfamilies of Traver; mayfly specialists in most of the world have generally recognized such groups as families for most of the present century. The family Behningiidae was unknown in 1935, and the family Oligoneuriidae was not known north of Mexico. The family Euthyplociidae was included then in Polymitarcyidae.

The arrangement of the subfamilies into families and superfamilies used in this book is indicated in Table 1, which lists all superfamilies and families. The groups in boldface type are not known to occur in North or Central America. A subfamily name is not listed if the family is not subdivided. Table 2 provides a comparison, in a general way, of the present classification of the North American forms with the earlier classifications of Burks (1953) and Needham, Traver, and Hsu (1935).

Use of the Keys

The keys offer only two mutually exclusive statements in each couplet, one of which should apply to the specimen being identified. At the end of each couplet the user is given either the number of the next pertinent couplet or the genus name of the specimen. The couplet numbers (except in the case of the first couplet) are followed by one or more numbers in parentheses indicating the previous couplet or couplets that led the user to that line. These backtrack numbers enable the user who suspects the identity of a genus to work through the key in reverse to verify the characters of the genus. The words ''in part'' indicate that a genus or a family is keyed out more than once; this is important when the user runs the key in reverse. At least one nymph is illustrated for each subfamily; in the half of any couplet in which a figured nymph is keyed out, reference is made to the figure number. These drawings accompany the text on the subfamily. The same procedure is followed for the adults, but there are very few such figures.

The sequence of key couplets is as follows. The first series of couplets (twenty-four in the nymphal keys and twenty-eight in the adult keys) cover the families and the subfamilies. The genera are keyed only when there is a single

---

Table 1. Families and Subfamilies of Ephemeroptera

Heptagenioidea	Tricorythidae
Siphlonuridae	**Tricorythinae**
Siphlonurinae	**Dicercomyzinae**
Rallidentinae	Leptohyphinae
Acanthametropodinae	**Machadorythinae**
Ameletopsinae	**Ephemerythinae**
Isonychiinae	Caenoidea
Coloburiscinae	Neoephemeridae
Oniscigastrinae	Caenidae
Metretopodidae	Prosopistomatoidea
Siphlaenigmatidae	Baetiscidae
Baetidae	**Prosopistomatidae**
Oligoneuriidae	Ephemeroidea
Chromarcyinae	Behningiidae
Oligoneuriinae	Potamanthidae
Heptageniidae	Euthyplociidae
Pseudironinae	Ephemeridae
Heptageniinae	Ephemerinae
Arthropleinae	Pentageniinae
Anepeorinae	Polymitarcyidae
Spinadinae	Polymitarcyinae
Ametropodidae	Campsurinae
Leptophlebioidea	**Asthenopodinae**
Leptophlebiidae	**Palingeniidae**
Ephemerellidae	
Teloganodinae	
Melanemerellinae	
Ephemerellinae	

NOTE: Bold type is used to indicate families and subfamilies which do not occur in North or Central America.

genus to key in the family or the subfamily. The genera of the various families are then keyed in accordance with the sequence of families in the text.

The keys are utilitarian. When necessary, we have keyed out a family or genus more than once to avoid resorting to difficult or obscure characters. The selection of key characters is often a matter of convenience, and difficult characters of significance may not be mentioned. For the diagnostic characters of any taxon, the descriptions under the appropriate accounts should be consulted. It should be understood that the characters chosen to key a genus are not necessarily those that are the basis for placing the forms in the genus. In some cases one stage is strongly differentiated and the other stage weakly differentiated; this sometimes makes it difficult to key the weakly differentiated stage. We have used the easiest and most reliable characters as the

Table 2. Comparison of the Three Classification Systems Most Used in North America

Edmunds, Jensen, and Berner (1976)	Burks (1953)	Needham, Traver, and Hsu (1935)
Siphlonuridae	Siphlonurinae[1]	Siphlonurinae[1]
Siphlonurinae	Siphlonurinae[1]	Siphlonurinae[1]
Acanthametropodinae	Ametropidae, part	(Not yet known)
Isonychiinae	Isonychiinae[1]	Siphlonurinae[1]
Baetidae	Baetinae[1]	Baetinae[1]
Oligoneuriidae	Oligoneuriidae	Oligoneuriinae[1]
Heptageniidae	Heptageniidae	Heptageniidae
Heptageniinae	Heptageniinae	Heptageniinae
Arthropleinae	Heptageniinae	Heptageniinae
Pseudironinae	Ametropidae, part	Metretopinae, part[1]
Anepeorinae	Heptageniidae	Heptageniinae[3]
Spinadinae	(Not yet known)	(Not yet known)
Ametropodidae	Ametropidae, part	Ametropinae[1]
Metretopodidae	Ametropidae, part	Metretopinae[1]
Leptophlebiidae	Leptophlebiidae	Leptophlebiinae[1]
Ephemerellidae	Ephemerellidae	Ephemerellinae[1]
Tricorythidae	Caenidae, part	Caeninae, part[1]
Neoephemeridae	Neoephemeridae	Neoephemerinae[2]
Caenidae	Caenidae, part	Caeninae, part[1]
Baetiscidae	Baetiscidae	Baetiscinae[1]
Behningiidae	(Not yet known)	(Not yet known)
Potamanthidae	Potamanthinae[2]	Potamanthinae[2]
Ephemeridae	Ephemerinae[2]	Ephemerinae[2]
Ephemerinae	Ephemerinae[2]	Ephemerinae[2]
Pentageniinae	Ephemerinae[2]	Ephemerinae[2]
Polymitarcyidae	Ephoroninae[2]	Ephoroninae[2]
Polymitarcyinae	Ephoroninae[2]	Ephoroninae[2]
Campsurinae	Campsurinae[2]	Campsurinae[2]

[1] Subfamily of Baetidae
[2] Subfamily of Ephemeridae
[3] Subfamily of Heptageniidae

first character pair in the couplet, but the secondary characters allow the keying of many incomplete specimens of these fragile insects. The use of secondary and tertiary key characters is frequently unnecessary if good specimens are available.

Persons unfamiliar with mayfly structure should consult the discussions of morphology and terminology and study the labeled drawings (figs. 20–25) in that section. The illustrations referred to in the keys are labeled only as necessary, but key characters are pointed out by arrows. The arrows are not labeled unless there are references to two or more structures on the same figure.

For each family one wing is illustrated with all the principal veins labeled and the diagnostic veins indicated by arrows. On other wing drawings only the veins cited in the key are labeled. The male genitalia are frequently diverse within a genus, and the species that is illustrated will often be quite different from the specimen being keyed.

Abdominal segments are referred to by number (segment 1 through segment 10). The segments are always counted from the last segment (segment 10) forward; this is because the first abdominal segment is frequently somewhat fused to the metathorax, which makes it difficult to count the segments accurately from front to back. All other labels on the drawings are cross-referenced to the keys.

Imago mayflies are usually necessary for species identification, but generic characters are present on the subimagoes. Most subimagoes can be distinguished by the usually dull translucent wings, the dull body surfaces, and the setae on the wing margins, but there are times when these features are inconclusive to the novice. Extremely small mayflies such as *Caenis* and *Tricorythodes* retain the setae on the wing margins, and in the Oligoneuriidae the subimaginal cuticle is retained on the wings of imagoes. The caudal filaments are usually covered with setae in subimagoes and are glabrous in the adult. Curiously, in the subimagoes of the Oligoneuriidae the caudal filaments are without fine setae and the imagoes have whorls of conspicuous setae. In some genera there is no imago stage in the females.

The nymphal key is for mature or semimature nymphs with well-developed wing pads. The keys will frequently be useful in the identification of young nymphs, but the younger the specimen, the less the likelihood of success. For example, most, if not all, first-instar nymphs have no gills, and their generic identity can be determined only when one is very familiar with mayfly nymphs and their age variation. The gills of young nymphs are on fewer segments and are usually narrower and of different form than are those of more mature nymphs. The claws also may differ between young and more mature nymphs. In the family Baetidae generic identifications are often difficult with mature nymphs, especially in genera from Mexico and Central America, and it is very difficult or even impossible to identify the young nymphs correctly.

In the keys to adults and nymphs the number of caudal filaments is utilized for identification. When the terminal filament is truly vestigial, it is represented by a small tapered rudiment of one or more segments. When a fully developed terminal filament has been broken off, the three basal remnants are of normal diameter.

In some cases, we have used the term *rare* in the keys. This term sometimes means no more than "infrequent in collections." The frequency of a species or a genus in collections represents the intersection rate of a prepared collector and the insect. If the nymphs of a fairly common species occur in deep, fast water where collecting is difficult, and if the adults are not attracted to light, or if they fly at dawn when most biologists are asleep, the species is "rare." In reality, there is a considerable range of abundance in various species and genera, but to translate this to abundance in collections one must bear in mind the seasonal and circadian rhythms and the behavior of both mayflies and mayfly workers.

Problem Couplets

Despite every effort to make the keys workable certain precautions are necessary, especially for the inexperienced biologist. We have tried to make the keys equally useful to professional entomologists and to fisheries biologists, fishermen, and limnologists who may have little knowledge of mayflies. Only time and use will tell us if we have whipped the bugaboo that "keys are made by biologists who don't need them for those who can't use them."

Keys to nymphs. Couplet 25 is rather complex because the nymphs of the siphlonurid genus *Ameletus* are frequently identified as Baetidae. If one learns to look for the distinctive maxillary crown of *Ameletus* and to note the general appearance of the genus, Siphlonuridae and Baetidae can be separated by the degree of development of the posterolateral projections on abdominal segments 8 and 9.

In connection with couplet 34 it should be noted that the dark bands on the caudal filaments often consist of large spines as in figure 84. The complex of couplets 35 through 38 can be bypassed if one can easily see that the hind wing pad is present (*Centroptilum*) or absent (*Cloeon*).

Couplets 64 through 68 are complex, but the common genus *Paraleptophlebia* must pass through this maze. It is our opinion that *Paraleptophlebia* is ancestral to *Habrophlebiodes* and that the peculiar reduced hind wings of *Habrophlebiodes* are related to its small size. Hence the nymphs of *Hab-*

rophlebiodes may prove to be inseparable from some species of *Paraleptophlebia*. The problem of the relationships of *Habrophlebiodes* should be solved by the studies on the entire Holarctic fauna of this complex now in progress by W. L. Peters and others. *Paraleptophlebia* is only superficially similar to *Thraulodes*, *Hagenulopsis*, *Homothraulus*, and *Hermanellopsis*; furthermore, the geographic overlap between *Paraleptophlebia* and *Thraulodes* or *Homothraulus* is small, and there is no known overlap of *Paraleptophlebia* with the two Central and South American genera (couplets 67–68).

Keys to adults. In the adult keys a number of couplets may be misinterpreted unless certain precautions are observed. In couplet 3 the first half of the couplet requires that *both veins* MP$_2$ and CuA diverge strongly from MP$_1$. Many of the mayflies that key in the first half of the couplet are among the largest of North American mayflies. Vein MP$_2$ *only* diverges in *Siphloplecton* in the second half of the couplet; this eastern North American genus has distinctive wing venation in MP and in the CuA–CuP region.

With regard to couplets 10 and 32, most of the species without hind wings are small (usually 7 mm long or less), and those with minute hind wings can be keyed to the correct family even if the hind wing is overlooked. The exceedingly small hind wings of females of a few small species of *Baetis* require careful examination to determine their presence.

Couplet 35 keys out *Dactylobaetis* very well if the expanded broad base for the costal projection is noted. We include a pair of auxiliary figures with a stippled arc on each wing which emphasizes the distinctive wing shapes used to separate *Baetis* from *Dactylobaetis*.

The four baetid genera keyed out in couplets 38 to 40 have not been distinguishable with certainty as adults. The keys and the figures should enable the four genera to be separated (but see the discussion of taxonomy under Baetidae). In Central America, Mexico, Texas, New Mexico, Arizona, and perhaps California there may be species that will not fit the characters we propose for the separation of these genera. The geography of the four genera should be used cautiously since the adults have been previously unidentifiable.

Couplets 50 to 56 represent a gallant attempt to key unassociated females of Heptageniidae to genus. Probably this part of the key will not work for all species. There are no published characters, and we have not seen female specimens which enable us to determine the key characters of female *Anepeorus* (subfamily Anepeorinae).

Couplets 59 and 60 may seem unnecessarily complex, but the females of *Leptophlebia johnsoni* are not separable from the females of the larger species of *Paraleptophlebia* on structure, hence the use of a superficial color character in an extra couplet for this species.

Morphology of Key Characters and Terminology

This section provides a brief account of the external morphology of mayflies to facilitate use of the keys. Individuals with little entomological training should study the figures in detail.

Characters of Adults

Head (fig. 20). The eyes in most mayfly species are sexually dimorphic. The eyes of the male are usually large and close together or touching dorsally on the vertex while those of the female are smaller and farther apart. In some species, such as most species of the families Caenidae and Tricorythidae, the eyes of both sexes are small and remote. The eyes of the male may be uniformly colored or the upper part may be of a different color, often orange to red. The eye facets may be uniform in size or the upper facets may be larger. In the Baetidae and some Leptophlebiidae the upper facets may be raised on a stalklike portion; such eyes are called turbinate or semiturbinate (fig. 233). The eyes of the female are generally uniform in color and facet size.

Anteroventral to the eyes are one median and two lateral ocelli and the antennae. The ocelli are variable in size and placement and may be somewhat elevated. The antennae, consisting of a short basal scape, a well-developed pedicel, and a slender filiform flagellum, usually lie somewhat lateroventral to the lateral ocelli. The antennae are usually shorter than the width of the head.

The frons make up the "face" of the mayfly. The shape of the ventral margin of the frons, called the frontal margin, is variable and is useful as a taxonomic character in some groups. The mouthparts in the adult are vestigial and nonfunctional.

Thorax (figs. 20, 21). The thorax consists of three regions, each with one or two pairs of appendages: the prothorax with the forelegs; the mesothorax with the middle pair of legs and the forewings; and the metathorax with the hind legs and hind wings (the latter absent in some forms).

The prothorax is usually small, and in males it may be partially concealed by the eyes. The shape of the posterior margin of the dorsal portion, the prothoracic notum, is used in separating several genera. The mesothorax and the metathorax are rather solidly fused into a large structure with the mesonotum making up most of the dorsal portion. Two indistinct sclerites occupy most of the mesonotum; these consist of a large convex anterior

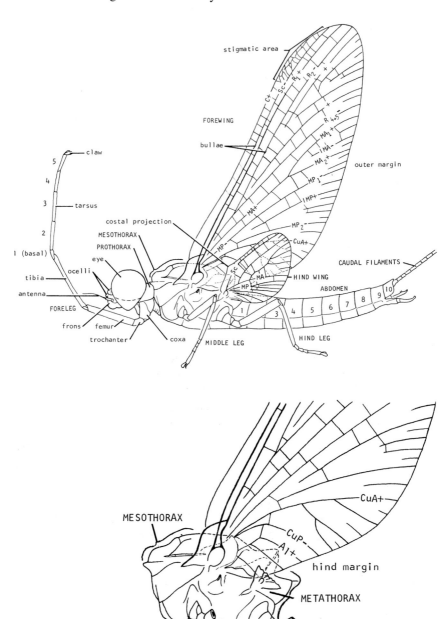

Figs. 20, 21. Adult, *Ephemerella*.

scutum and a smaller posterior scutellum. The shape of the scutellum is used for separating some genera of Baetidae and Ephemeridae.

Legs (fig. 20). Each leg consists of six parts: a rather stout basal coxa; a small trochanter; a large and somewhat flattened femur; a slender cylindrical tibia; a slender cylindrical tarsus with four or five segments (the basal segment fused in varying degrees to the tibia) and apical claws. The forelegs of most mayflies show sexual dimorphism, with those of the male having very long tibiae and tarsi and usually being much longer than the middle and hind legs (often as long as or longer than the body). The forelegs of the female are about the same size and proportion as the middle and hind legs. In the Polymitarcyidae the middle and hind legs of the male and all legs of the female are vestigial and presumably nonfunctional, and in *Dolania* (Behningiidae) all the legs of both sexes are vestigial but apparently are somewhat functional.

The number of segments in the hind tarsi and the relative proportions of the fore and hind tarsal segments are used taxonomically. The claws are variable — both sharply pointed, or one pointed and one blunt, or both blunt.

Wings (figs. 20, 21). Most mayflies have two pairs of wings; the forewings arise from the mesothorax, and the much smaller hind wings arise from the metathorax. In the families Caenidae, Tricorythidae, Baetidae, and some Leptophlebiidae the hind wings have become greatly reduced or completely lost.

The forewings are somewhat triangular in shape; the three margins are known as the costal margin, the outer margin, and the hind margin. The surface of the wing has a regular series of corrugations or fluting with the longitudinal veins lying either on a ridge (indicated in the figures and elsewhere by +) or a furrow (indicated by −). Weakened areas called bullae occur on some of the anterior longitudinal furrow veins and allow the wings to bend during flight.

Venational nomenclature is not consistent among various authors. We follow the system proposed by Tillyard (1932) as discussed by Edmunds and Traver (1954). The abbreviations used for designating the major longitudinal veins and their convexity (+) or concavity (−) relative to wing fluting are as follows:

C (+) = costa
Sc (−) = subcosta
R_1 (+), R_2 (−), R_3 (−), R_{4+5} (−) = radius$_1$, radius$_2$, etc.
MA_1 (+), MA_2 (+) = medius anterior$_1$, etc.
MP_1 (−), MP_2 (−) = medius posterior$_1$, etc.
CuA (+) = cubitus anterior
CuP (−) = cubitus posterior
A_1 (+) = anal$_1$

R_2 through R_5 are often referred to as the radial sector (Rs). The intercalary veins lie between the principal veins. When the intercalaries are long, they lie opposite the principal veins on either side; for example, IMA (intercalary medius anterior) is a furrow $(-)$ vein lying between the two ridge $(+)$ branches of MA. Conversely, IMP is a ridge vein $(+)$ in between the furrow veins of MP. The longer veins alternate as ridges and furrows at the wing margin.

Several areas of the forewings are referred to in the keys. The stigmatic area is the rather short region at the apex between the costal and subcostal veins. The disc is the central area of the wings. The cubital region lies between the cubitus anterior and cubitus posterior veins. Two ridge veins of the forewing, MA and CuA, are important landmarks in locating and identifying the entire venation. Learning which veins are ridge veins $(+)$ and which are furrow veins $(-)$ is well rewarded in time saved in using the keys.

The venation of the hind wing is often difficult or impossible to interpret, but only three of the veins (subcosta, medius anterior, and medius posterior) are used to any extent in the keys. The presence or absence of the hind wings and the shape and the nature of the costal projection (if present) are frequently used characteristics.

Abdomen (fig. 20). Except for a few species such as *Siphlonisca aerodromia* and *Ephemerella (Timpanoga) hecuba*, the abdomens of most mayflies are essentially the same and consist of ten segments. Each segment is ring-shaped and consists of a dorsal tergum and a ventral sternum.

The posterior portion of sternum 9 of males (fig. 22) is called the subgenital plate (or sometimes the styliger plate). The posterior margin of the subgenital plate, which is variable in shape, gives rise to a pair of slender and usually segmented appendages called the forceps (or the claspers). In some genera (*Caenis*, *Tortopus*, and *Campsurus*) the forceps consist of a single segment; in others they consist of two, three, or more segments. Dorsal to the subgenital plate are the paired penes; these are often fused to some degree (fig. 22). In most species the penes are well developed and easily observed; in the Baetidae they are membranous and extrudable. A variety of lobes, spines, and processes of various shapes are found on the penes of many species and provide useful taxonomic characters.

The posterior portion of sternite 9 of the female is referred to as the subanal plate. The posterior margin is variable in shape and may be used to distinguish some genera.

Arising from the posterior portion of tergum 10 are the caudal filaments (figs. 20, 23). Most species have two caudal filaments, the cerci, and a vestige of a median terminal filament; others have three caudal filaments, the cerci, and a median terminal filament. The caudal filaments are variable in length, but most are two to three times as long as the body. The terminal filament may be equal in length and thickness to the cerci or shorter and more slender than the cerci. In the subgenus *Caudatella* of *Ephemerella* the cerci are shorter than the terminal filament.

Characters of Mature Nymphs

Head (fig. 24). The shape of the head is variable among the genera and may possess a variety of processes, projections, and armature. The eyes are usually moderately large and are situated laterally or dorsally near the posterolateral margin. A median ocellus and a pair of lateral ocelli are usually located between the eyes. A pair of filiform antennae usually arises anterior or ventral to the eyes. The antennae may be short or less than the width of the head to more than twice as long as the width of the head.

Mouthparts (fig. 25). Unlike the adults, the nymphs have fully functional mouthparts which are adapted according to feeding habits. The mouthparts are usually concealed beneath or behind the head capsule (although portions may be exposed). In nymphs with hypognathous heads the mouthparts are oriented vertically and are directed ventrally. In nymphs with prognathous heads the mouthparts are oriented horizontally to the substrate and are directed an-

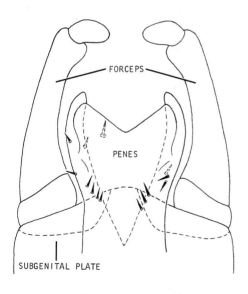

Fig. 22. Genitalia (adult male), *Ephemerella*.

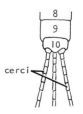

Fig. 23. Apex of abdomen (adult female), *Ephemerella*.

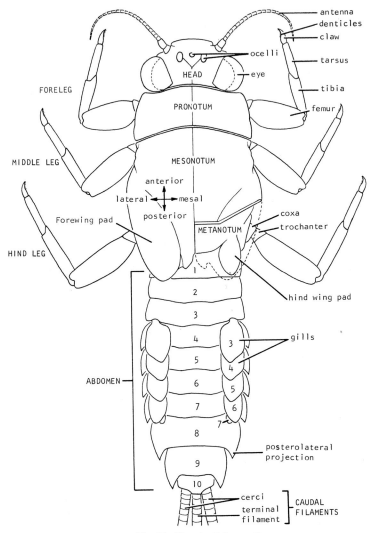

Fig. 24. Nymph, *Ephemerella*.

teriorly. In such cases the face is still regarded as anterior and the underside is regarded as posterior. The most anterior mouthpart is the labrum, which articulates with the clypeus of the head capsule. The labrum may be as wide as or wider than the head capsule, or it may be very small and narrow.

The mandibles lie posterior to the labrum. The incisors of the left mandible are usually similar in shape to those of the right mandible. The molar surfaces

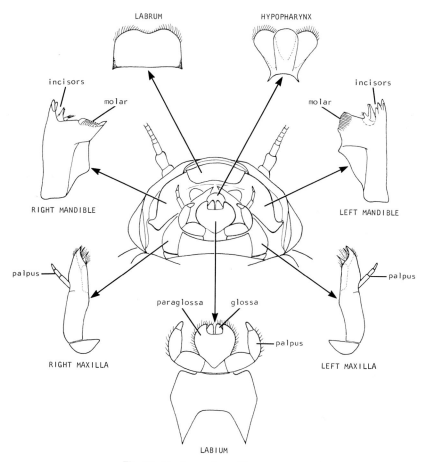

Fig. 25. Mouthparts (nymph), *Ephemerella*.

of the two mandibles usually differ: on the left mandible, the molar surface is oriented somewhat parallel to the lateral margin of the mandible; the molar surface of the right mandible is oriented at a right angle to the lateral margin. Both the incisors and the molar surfaces are heavily sclerotized. Behind the mandibles is a pair of maxillae. The anterior portion of each maxilla consists of the galea-lacinia (or galeolacinia); the posterior portion consists of the stipes and the cardo. Arising from the lateral surface of each galea-lacinia is a two- or three-segmented palpus (rarely absent). Next is the hypopharynx, which usually consists of two structures: the median lingua (or hypopharynx proper) and the lateral superlingua (or parapsides). The superlingua may be absent, only slightly developed, or highly developed and expanded laterally.

The most posterior mouthpart is the labium. The main body of the labium consists of a small submentum (postmentum) and a larger mentum (prementum). Arising from the mentum are two pairs of lobes: the glossae (the mesal pair) and the paraglossae (the lateral pair). Also arising from the mentum of the labium is a pair of palpi. Each palpus may consist of two or three segments.

Thorax (fig. 24). Three thoracic segments are present (the prothorax, the mesothorax, and the metathorax), each bearing a pair of legs. The developing adult wings in the wing pads are also attached to the mesothorax and the metathorax.

The dorsal surface of the prothorax, the pronotum, is better developed than in the adults and may bear a variety of projections. The largest part of the thorax is the mesonotum with the forewing pads. In mature nymphs the wing pads extend from the pronotum posteriorly to abdominal segment 1 or 2 or beyond. In *Baetisca* the mesonotum forms a shieldlike carapace extending to about the middle of abdominal segment 6. The metanotum is usually concealed dorsally by the mesonotum. It is much smaller than the mesonotum and, except in those species whose adults lack hind wings, bears the hind wing pads.

The thoracic sterna in most species are small and rectangular. In a few genera they have well-defined ridges or spines.

Legs (fig. 24). Each leg consists of a stout coxa, a small trochanter, a more or less flattened femur, a cylindrical or subtriangular tibia, a cylindrical unsegmented tarsus, and a single claw usually armed with denticles. The legs of the nymphs are shorter and stouter than those of the adult, and the hind legs are usually longer than the forelegs. There is great variety in the structure of the legs; for instance, they may have various arrangements of spines, tubercles, and/or setae. The legs of some genera are modified for such special functions as burrowing, filtering food, grooming, and gill protection.

Abdomen (fig. 24). All mayflies have ten-segmented abdomens, although some of the segments may be concealed beneath the mesonotum. The abdominal terga may have spines and/or tubercles, and the shape of the posterolateral corners of each of the terga may provide good taxonomic characters. The sterna are usually free of armature, although in some species conspicuous spines or setae are present.

Gills (fig. 24). The gills are the most variable structures to be found in mayfly nymphs. In most mayflies they are located entirely on the abdomen, although in some species they may also occur at the bases of the coxae (*Isonychia*, *Dactylobaetis*) or at the bases of the maxillae (*Isonychia*, Oligoneuriidae). Gill position on the abdomen is also variable. In Oligoneuriidae the gills on segment 1 are ventral, and those on succeeding segments are dorsal; in *Dolania* and *Anepeorus* the gills are ventral, although in most mayflies they are lateral or dorsal. The gills may occur on abdominal segments 1 through 7 or may be absent from one or more segments in various combinations. They may be absent from either or both ends of the series. In *Ephemerella* they are always absent from segment 2 and sometimes from segment 3, but vestigial gills may occur on segment 1.

The structure of the gills is highly variable. Those of the middle abdominal segments may each consist of two platelike lamellae which may be of various shapes; two lamellae with the margins fringed or finely dissected; a dorsal platelike lamella and a ventral fibrilliform tuft; a dorsal platelike lamella with the ventral lamella dissected into a number of small lobes; a single platelike lamella; or other configurations (the illustrations show the full range of variability). Gills on abdominal segment 1, when present, may be rudimentary, or similar in shape to those of succeeding pairs of gills but usually smaller, or different in shape in comparison to succeeding pairs. Gills on abdominal segment 7 are usually similar in shape to the preceding gills but smaller.

Caudal filaments (fig. 24). Most species have three caudal filaments, a terminal filament, and two cerci. In some species only the cerci are well developed, the terminal filament being represented by a short rudiment or entirely absent. As in the adults, the terminal filament, when present, may vary in length and thickness relative to the cerci. The length of the caudal filaments varies from shorter than the body to two or three times the length of the body.

Terminology

The terminology applied to immature mayflies is the source of considerable confusion. British workers tend to call all mayfly immatures *nymphs*, although some refer to them as *larvae*. Most continental European workers refer to mayfly immatures as *larvae*, and some call immatures in the growth stages *larvae* and those in the metamorphosing stages *nymphs*. Very young mayfly immatures are sometimes referred to as *nymphules*. The stages between molts of the exoskeleton are called *instars*, but this term is also used by some workers to designate size groups.

In this book we have used the term *nymph* for all stages of immature mayflies; this term has wide usage and is preferred by two of the three authors of this book. Edmunds believes the term *larva* should be used for mayfly immatures, in conformity with usage by many European entomologists and most other zoologists. The restriction of the word *larva* to the immatures of insects with complete metamorphosis by entomologists is not consistent with widespread use of the same word for the immatures of any animals that undergo metamorphosis. The problem of terminology can be resolved partially by assuming that the term *nymph* is just one of the many terms applied to various types of larvae.

In this book (especially in the keys) a conscious but not always successful attempt has been made to reduce specialized terminology. For example, in reference to gills, we have used the common term *fringed* rather than *fimbriate*, which means the same thing. However, we use such terms as *cordate* and *obovate* when they are shorter or more precise than common terms, assuming that the drawings make the meaning clear. Most of these terms are defined in unabridged dictionaries, and the others can be found in specialized entomological or zoological dictionaries or glossaries.

Distribution

In the key to genera we have cited distributions in a way which should be useful for interpreting the most likely areas of ecological occurrence of any genus and most nearly predictive of the unrecorded range of occurrence. From our best judgment on the basis of the known distribution of each genus and from our estimate of the closest cladistic relatives of each genus (those from which it has branched most recently), we indicate for each genus whether it is of austral or boreal origin.

The known distribution and the known ecological requirements allow us to apply approximate limits to the distribution of the genera. The meaning of distributions stated in these terms is greater than the mere statement of known distribution. For example, on the basis of cladistic relationships we believe that the genus *Paracloeodes* is of austral origin, and we have assigned it as mid-austral. Its known distribution is in California, Wyoming, Minnesota, Mississippi, Alabama, Georgia, and Puerto Rico, but it would not be surprising to discover new records from warm rivers in the United States north and east of the areas of known distribution. The genus may also occur in Saskatchewan. Further, its occurrence in Mexico seems highly likely, and it probably formerly occurred in some of the now polluted rivers of the central United States. The mid-austral genus *Traverella*, whose distribution is much better known than *Paracloeodes*, reaches its known northern limits in the somewhat silted Saskatchewan River and probably does not occur much farther north than present records indicate. Austral forms are expected to reach their northern limits at low elevations in valley streams that are frequently warmer than other streams because of their silt load. The same predictive pattern holds for boreal genera that reach their southern limits at high elevations or near the source of cold spring-fed streams. The known southern limits of the upper boreal genus *Parameletus* are in north-central Utah at high altitudes where the nymphs develop rapidly and emerge early in the season. However, it would not be unexpected to find *Parameletus* at high altitudes as far south as the higher peaks of the Rocky Mountains in northern New Mexico. Early emergence is frequently a characteristic of boreal forms in the southern part of their range or at low elevations. A few boreal forms develop in the autumn and emerge in October, November, or December, even in the relatively cold winters in Utah.

Although the placement of most genera has been done with considerable confidence, the assignment of some genera as either austral or boreal is less certain. The genus *Hexagenia* may be boreal in its affinities rather than austral. If the subgenus *Hexagenia* was derived from the subgenus *Pseudeatonica* of South America and Central America, it represents the only case of subgeneric differentiation on the two New World continents. In contrast, *Hexagenia* has no closely related Eurasian counterparts, which would suggest a boreal origin. *Pentagenia* appears to be an ancient endemic and its origin is not certain, thus its assignment as austral is somewhat arbitrary. The genus *Litobrancha* also may be an old endemic, but it probably is of boreal origin. The endemic family Baetiscidae (*Baetisca*) probably evolved in North America; the most closely related family, Prosopistomatidae, appears to have evolved in part of Gondwanaland (Africa, Madagascar, and India). Although *Choroterpes* clearly has many species of boreal affinities, the possibility of an austral origin for part of the genus cannot be ruled out. The genus *Baetis*, as currently recognized, almost certainly has boreal and austral elements. Further studies will probably clarify the distributional relationships of some of these genera.

The subdivision of the boreal and austral groups into upper, middle, and lower subgroups is somewhat arbitrary, but generally it corresponds to major patterns in nature. For boreal groups the keys usually mention only the southern limits, and for austral groups they usually cite only the northern limits. The various distribution patterns used in the keys are defined in the following section.

Austral Distribution

The austral genera are those we believe to have originated in South America or from South American ancestors. Many such forms are either South American or African (primarily pantropical) or are allied to largely endemic South American groups.

The term *lower austral* is applied to genera that are primarily obligates of tropical waters; such forms are expected to reach their northern limits in southern Mexico, although some may intrude farther north in warm rivers in limited areas.

The *mid-austral* forms generally extend through Central America and Mexico into the southern United States and frequently into the central United States and central Canada, having dispersed there via the Mississippi and

Missouri river system or the Colorado river system. Most northern species of groups of austral origin are not necessarily adapted for cooler waters; in Utah most of the mid-austral species hatch from the egg only after the water warms up to almost 70° F and emerge as adults before the water cools in the autumn. A few genera that seem to have mid-austral patterns are as yet unknown in Mexico, and we may have misjudged their affinities and origins.

The few forms that we call *upper austral* seem to us to be quite clearly of southern origin, but a few species extend far north into the cool ponds and streams of Canada.

Boreal Distribution

The boreal genera are of Holarctic distribution or are Nearctic but closely cladistically related to Holarctic genera.

The *upper boreal* forms are generally confined to Canada and the northern (often montane) United States. In western North America such groups extend south down the Sierra Nevada; in the Rocky Mountains they seldom occur south of the Snake River plains of Idaho and the high plains of southern Wyoming. In the eastern United States such genera seem largely limited to Minnesota, Michigan, and upper New England.

The *mid-boreal* forms are generally distributed farther south in the United States, but they are largely adapted to cool water. In some cases a very few species extend into montane Mexico or even into montane Central America. The route to the south appears to have been along the Sierra Nevada and down the Mexican and Central American Cordillera. Most such groups extend to the mountains of Arizona and New Mexico and down the Appalachians to North Carolina and Tennessee and frequently even into Florida. In the central United States they are found as far south as the Ozark area. We have also included here those boreal genera that seem to be adapted to warm rivers but are not known to occur in Mexico or Central America. Almost certainly some of these genera occur in Mexico but are as yet unknown there.

The *lower boreal* forms are those with some species adapted for warmer rivers and with a range extending into Mexico and sometimes into South America.

Keys to Families and Genera

Geographic subdivisions used in discussions and charts of species distribution.

Key to Nymphs

1   Thoracic notum enlarged to form a shield extended to abdominal segment 6, gills enclosed beneath shield (fig. 26); nymph shown (fig. 423); mid-boreal, widely distributed in eastern and central North America, rare in west . . . . Baetiscidae, *Baetisca* (p. 269)

       Thoracic notum not enlarged as above; abdominal gills exposed
. 2

2(1)   Mandibles with large tusks (t) projected forward and visible from above head (figs. 30–34) 3

       Mandibles without such tusks (figs. 42, 53, 56, 135–138) . . .9

3(2)   Gills on abdominal segments 2–7 forked, without margins fringed (fig. 29); mid-boreal, western North America south to California and Utah . . . .
. . . . Leptophlebiidae, *Paraleptophlebia*, in part (p. 230)

       Gills on abdominal segments 2–7 forked or bifid, with margins fringed (figs. 27, 28) 4

4(3)   Fore tibiae and tarsi more or less flattened, adapted for burrowing (figs. 30, 33, 34); abdominal gills held dorsally 5

       Fore tibiae and tarsi cylindrical, unmodified (fig. 32); abdominal gills held laterally 8

5(4)   Ventral apex of hind tibiae projected into distinct acute point (fig. 35); mandibular tusks (t) curved upward apically as viewed laterally (fig. 37) . . . . Ephemeridae, 6

       Ventral apex of hind tibiae rounded (fig. 36); mandibular tusks (t) curved downward apically as viewed laterally (fig. 38)
. . . . Polymitarcyidae, 7

6(5)   Mandibular tusks (t) with a distinct dorsolateral keel which is more or less toothed (figs. 30, 41), with a line of spurs along the toothed edge; nymph shown (fig. 430); mid-austral, central North America, Texas, and Florida north to Manitoba . . . .
. . . . .Pentageniinae, *Pentagenia* (p. 292)

Mandibular tusks (t) more or less circular in cross section, without a distinct toothed keel (fig. 37) . . . . .Ephemerinae, 82

7(5)   Mandibular tusks with numerous tubercles (tb) on upper surface (fig. 38); fore tarsi unmodified, not partially fused with tibiae (fig. 40); nymph shown (fig. 431); mid-boreal, widespread in southern Canada and northern United States south to Georgia, Ohio, Kansas, and New Mexico . . . .
. . . . .Polymitarcyinae, *Ephoron* (p. 295)

Mandibular tusks with one or two prominent tubercles (tb) on inner margin (fig. 33); fore tarsi broad and partially fused to tibiae (fig. 39) . . . . .Campsurinae, 85

8(4)   Mandibular tusks (t) with numerous distinct long setae (fig. 43); caudal filaments with only inconspicuous setae; lower austral, north to Vera Cruz (Mexico) . . . . .Euthyplociidae, 81

Mandibular tusks (t) with only inconspicuous setae (fig. 32); caudal filaments with distinct row of setae on both sides, at least in apical half; nymph shown (fig. 425); mid-boreal, eastern North America west to Nebraska and Kansas . . . .
. . . . .Potamanthidae, *Potamanthus* (p. 276)

9(2)   Head and prothorax with dorsal pad of long spines on each side (fig. 42); gills ventral, those on segments 2–7 bifid with margins fringed; nymph shown (fig. 424); rare; mid-boreal, southeastern United States south to Florida . . . .
. . . . .Behningiidae, *Dolania* (p. 274)

Head and prothorax without pad of spines (figs. 53, 56, 128, 167); gills variable, not as above 10

10(9)   Forelegs with double row of long setae (s) on inner surface (figs.

NYMPHAL CHARACTERS. Fig. 26. *Baetisca escambiensis* (legs removed). Fig. 27. Gill 4, *Hexagenia limbata*. Fig. 28. Gill 4, *Ephoron album*. Fig. 29. Gill 4, *Paraleptophlebia bicornuta*. Fig. 30. Head and foreleg, *Pentagenia vittigera*. Fig. 31. Head and forelegs, *Paraleptophlebia packi*. Fig. 32. Head and foreleg, *Potamanthus* sp. Fig. 33. Head and foreleg, *Tortopus* sp.

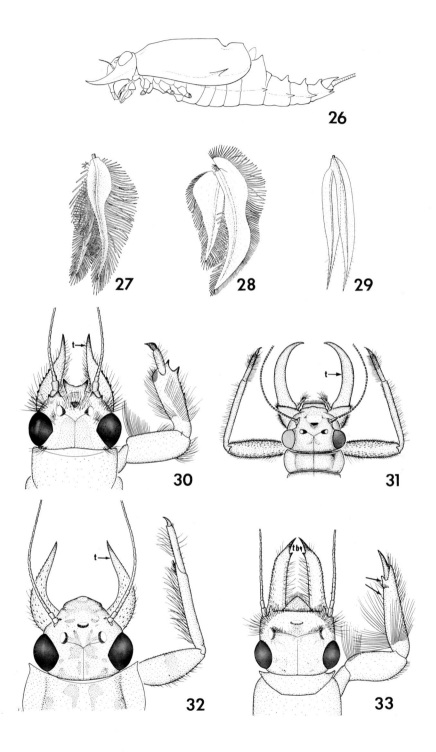

44, 45); tufts of gills (gm) present at bases of maxillae (fig. 44); gills (gt) may be present at bases of fore coxae (fig. 45) ...11

Forelegs with setae other than above (figs. 57a, 58a); gill tufts absent from bases of maxillae and fore coxae 12

11(10)   Gills (g₁) ventral on abdominal segment 1 (fig. 44); gill tufts absent from bases of fore coxae ....Oligoneuriidae, 47

Gills dorsal on abdominal segment 1; gill tufts (gt) present at bases of fore coxae (fig. 45); nymph shown (fig. 394); lower boreal, southern Canada south to Honduras ....
............Siphlonuridae, Isonychiinae, *Isonychia* (p. 144)

12(10)   Gills (g₂) on abdominal segment 2 operculate or semioperculate, covering succeeding pairs (fig. 47) 13

Gills (g₂) on abdominal segment 2 neither operculate nor semioperculate, either similar to those on succeeding segments or absent (figs. 50, 117).............................15

13(12)   Gills (g₂) on abdominal segment 2 triangular, semitriangular, or oval, not meeting medially (fig. 47); gill lamellae on segments 3–6 simple or bilobed, without margins fringed (fig. 51) ....Tricorythidae, 79

Gills (g₂) on abdominal segment 2 quadrate, meeting or almost meeting medially (fig. 46a); gill lamellae on segments 3–6 with margins fringed (fig. 46b) 14

14(13)   Mesonotum with distinct rounded lobe on anterolateral corners (fig. 48); operculate gills fused medially; developing hind wing pads present; nymph shown (fig. 420); rare, except locally in southeastern United States; mid-boreal, southeastern Canada west to Michigan and south to Florida ....
.................Neoephemeridae, *Neoephemera* (p. 261)

Mesonotum without anterolateral lobes (fig. 49); operculate gills not fused medially; developing hind wing pads absent ....
....Caenidae, 80

NYMPHAL CHARACTERS. Fig. 34. Head and foreleg, *Ephoron album*. Fig. 35. Hind leg, *Hexagenia*. Fig. 36. Hind leg, *Ephoron album*. Fig. 37. Head, *Ephemera simulans*. Fig. 38. Head, *Ephoron album*. Fig. 39. Foreleg, *Campsurus*. Fig. 40. Foreleg, *Ephoron album*. Fig. 41. Mandibular spur variations, *Pentagenia vittigera*. Fig. 42. Head, *Dolania americana*. (Fig. 42 from Edmunds and Traver, 1959.)

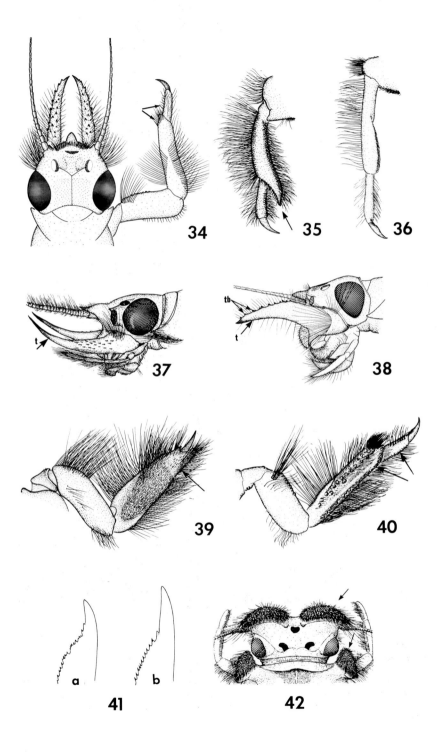

34 35 36

37 38

39 40

41 42

15(12)   Gills absent on abdominal segment 2, rudimentary or absent on segment 1, and present or absent on segment 3 (figs. 50, 167, 183); gills on segments 3–7 or 4–7 consist of anterior (dorsal) oval lamella and posterior (ventral) lamella with numerous lobes (fig. 52); paired tubercles often present on abdominal terga (fig. 50); mid-boreal, widespread in Canada and United States south to northern Sonora and Baja California (Mexico) . . . .
. . . . .Ephemerellidae, *Ephemerella*, 69

Gills present on abdominal segments 1–5, 1–7, or 2–7; paired tubercles rarely present on abdominal terga 16

16(15)   Nymph distinctly flattened; head prognathous, eyes and antennae dorsal (figs. 53, 56, 128). [Note: Some Leptophlebiidae appear to be intermediate in flattening, but either half of the couplet leads to Leptophlebiidae.] 17

Nymph not flattened, or only slightly flattened, being more cylindrical; head hypognathous, eyes and/or antennae lateral, anterolateral, or on front of head (figs. 78, 111, 112, 138) . . .
. 22

17(16)   Abdominal gills of single lamellae, usually with fibrilliform tufts (f) at or near bases (figs. 59, 129, 131–132), rarely pointed (figs. 54, 55) and rarely with narrow lanceolate branch making gills appear forked (fig. 55); mandibles concealed beneath flattened head capsule (figs. 53, 128); labial palpi two-segmented . . . . .
. . . . .Heptageniidae, 18

Abdominal gills either forked (figs. 148–154), formed of two lamellae with margins fringed (figs. 139, 141, 142) or terminated in filaments or points (figs. 140, 144b, 145, 146b, 147b), never formed of a lamella and fibrilliform tuft; mandibles (md) visible and forming part of upper surface of head (fig. 135); labial palpi three-segmented . . . . .Leptophlebiidae, in part, 57

18(17)   Maxillary palpi very long, adapted as sweeping structures with long conspicuous setae (fig. 56); nymph shown (fig. 405); upper

---

NYMPHAL CHARACTERS. Fig. 43. Head, *Campylocia* sp. Fig. 44. Ventral anterior portion, *Lachlania powelli*. Fig. 45. Foreleg, *Isonychia* sp. Fig. 46. Gills 2 and 4, *Caenis* sp. Fig. 47. Abdomen, *Tricorythodes minitus*. Fig. 48. Head and thorax, *Neoephemera youngi*. Fig. 49. Head and thorax, *Caenis simulans*. Fig. 50. Abdomen, *Ephemerella (Serratella) tibialis*. Fig. 51. Gill 4, *Tricorythodes minutus*. Fig. 52. Gill, segment 3, lamella raised, *Ephemerella (Drunella) grandis*. Fig. 53. Head, *Heptagenia* sp. (Figs. 44–46, 53 from Edmunds, Allen, and Peters, 1963.)

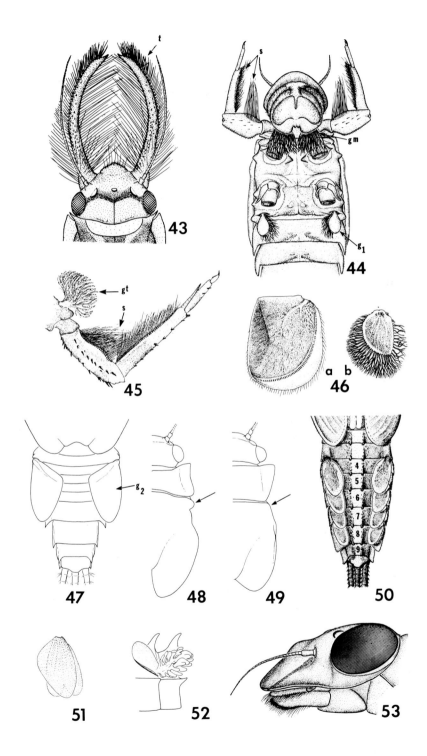

boreal, northeastern North America south to Wisconsin, Ohio, Quebec, and Massachusetts . . . .

. . . . .Arthropleinae, *Arthroplea* (p. 206)

Maxillary palpi much shorter, not adapted as above (fig. 128) .

. 19

19(18)   Claws elongate, as long as or longer than tarsi; gills narrow with lanceolate branch arising near middle (fig. 55); nymph shown (fig. 406); rare; mid-boreal, central and southeastern North America west to Manitoba, Wyoming, and Utah . . . .

. . . . .Pseudironinae, *Pseudiron* (p. 208)

Claws much shorter than tarsi; gills variable, not as above . .20

20(19)   Abdominal gills all inserted dorsally or laterally, but gills on segments 1 (g_1) and 7 (g_7) sometimes extended ventrally; lamellae on segments 2–6 or 2–7 broad (figs. 117, 118, 120)

. . . . .Heptageniinae, 48

Abdominal gills on segments 2–3 inserted ventrally, lamellae (lm) on segments 2–3 slender, about same length as fibrilliform portion (f) (as in fig. 59) 21

21(20)   Gills on all abdominal segments similar in shape and position; without dorsal tubercles on head, thorax, and abdomen; nymph shown (fig. 407); rare; mid-boreal, central North America southeast to Georgia and west to Utah . . . .

. . . . .Anepeorinae, *Anepeorus* (p. 210)

Gills on abdominal segments vary in form and position; dorsal tubercles present on head, thorax, and abdomen; nymph shown (fig. 408); rare; mid-boreal, Wisconsin, Indiana, and Georgia . . . . .Spinadinae, *Spinadis* (p. 212)

22(16)   Claws (cl) of forelegs differ in structure from those on middle and hind legs; claws of middle and hind legs long and slender, about as long as tibiae (figs. 57, 58) 23

NYMPHAL CHARACTERS. Fig. 54. Gill 4, *Arthroplea bipunctata*. Fig. 55. Gill 4, *Pseudiron* sp. Fig. 56. Head, *Arthroplea bipunctata*. Fig. 57. (a) Foreleg, (b) hind leg, *Ametropus albrighti*. Fig. 58. (a) Foreleg, (b) hind leg, *Siphloplecton basale*. Fig. 59. Abdomen, lateral, *Anepeorus* sp. Fig. 60. Head *Baetis*. Fig. 61. Head, *Siphlonurus*. Fig. 62. Apex of abdomen, *Siphlonurus*. Fig. 63. Apex of abdomen, *Ameletus*. Fig. 64. Apex of abdomen, *Callibaetis* sp. Fig. 65. Apex of abdomen, *Callibaetis* sp. Fig. 66. Apex of abdomen, *Cloeon*. (Figs. 56, 57 from Edmunds, Allen, and Peters, 1963.)

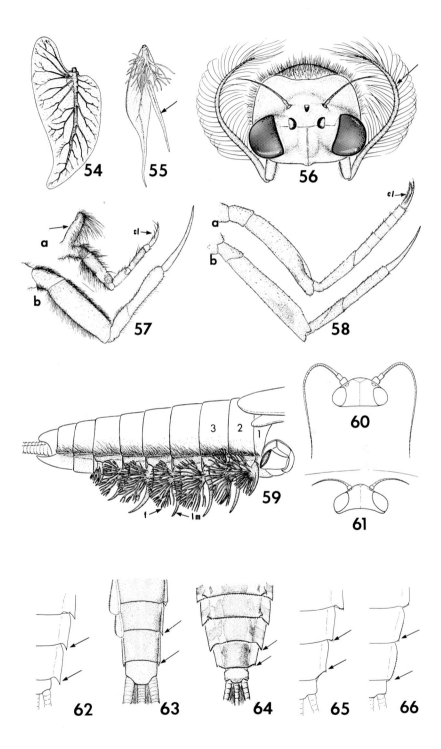

54 55 56 57 58 59 60 61 62 63 64 65 66

Claws (cl) of all legs similar in structure, usually sharply pointed, rarely spatulate; claws variable in length, if those of middle and hind legs long and slender, then usually shorter than tibiae (longer than tibiae in three rare genera) (figs. 70, 73) . . .
. 24

23(22)   Claws (cl) on forelegs simple, with long slender denticles; spinous pad present on fore coxae (fig. 57); nymph shown (fig. 396); mid-boreal, western North America south to Oregon, Utah, and New Mexico . . . . .Ametropodidae, *Ametropus* (p. 152)

Claws (cl) on forelegs bifid; without spinous pad on fore coxae (fig. 58) . . . . .Metretopodidae, 56

24(22)   Abdominal gills on segments 2–7 either forked (figs. 148–154), in tufts (fig. 143), with all margins fringed (figs. 141, 142), or with double lamellae terminated in filaments or points (figs. 144b, 145, 146b, 147b) . . . . .Leptophlebiidae, in part, 57

Abdominal gills not as above; gills either obovate (figs. 76, 77), cordate, or subcordate (fig. 79), lamellae either single, double, or triple (figs. 77, 74, 75, 71), never terminating in filaments or points; inner margin of gills usually entire, rarely finely dissected (fig. 69) 25

25(24)   Antennae short, length less than twice width of head (fig. 61); posterolateral projections present and usually prominent on abdominal segments 8–9 (fig. 62) but only moderately developed in *Ameletus* (fig. 63) in which gills (figs. 76, 77) and pectinate spines (ps) on crown of maxillae (fig. 78) are diagnostic; glossae (gs) and paraglossae (pg) of labium short and broad (fig. 67) . . . . .Siphlonuridae, 26

Antennae long, length more than three times width of head (fig. 60); posterolateral projections usually absent or small to moderately developed on abdominal segments 8–9 (figs. 65, 66, 64); glossae (gs) and paraglossae (pg) of labium long and narrow (fig. 68) . . . . .Baetidae, 32

NYMPHAL CHARACTERS. Fig. 67. Labium, *Siphlonurus* sp. Fig. 68. Labium, *Baetis bicaudatus*. Fig. 69. Gill 4, *Acanthametropus pecatonica*. Fig. 70. (a) Foreleg, (b) middle leg, (c) hind leg, *Analetris eximia*. Fig. 71. Gill 4, *Analetris eximia*. Fig. 72. (a) Foreleg, (b) middle leg, (c) hind leg, *Siphlonurus* sp. Fig. 73. (a) Foreleg, (b) middle leg, (c) hind leg, *Edmundsius agilis*. Fig. 74. Gill 2, *Edmundsius agilis*. Fig. 75. Gill 2, *Siphlonurus* sp. Fig. 76. Gill 4, *Ameletus* sp. Fig. 77. Gill 4, *Ameletus* sp. (Figs. 70, 71 from Edmunds and Koss, 1972.)

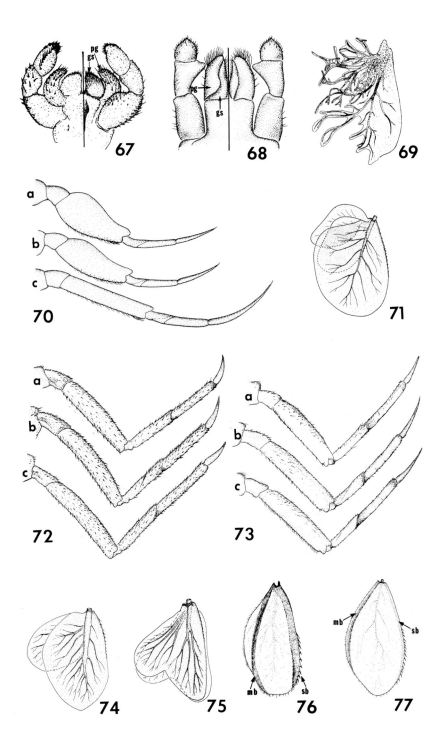

26(25)   Claws long and thin, much longer than tarsi of hind legs, tibiae and tarsi bowed (fig. 70); rare . . . . . .Acanthametropodinae, 27

Claws shorter than tarsi of all legs, tibiae and tarsi straight (figs. 72, 73) . . . . .Siphlonurinae, 28

27(26)   Abdominal terga each with small tubercle near the center; inner margin of gills finely dissected (fig. 69); rare; mid-boreal, Illinois and South Carolina . . . .*Acanthametropus* (p. 141)

Abdominal terga without tubercles; inner margin of anterior (dorsal) lamellae of gills entire, two smaller posterior (ventral) flaps present (fig. 71); nymph shown (fig. 393); rare; mid-boreal, Utah, Wyoming, and Saskatchewan . . . .*Analetris* (p. 142)

28(26)   Abdominal gills with double lamellae on segments 1–2 (in some species double also on segments 3–7) (figs. 74, 75) 29

Gills on all abdominal segments with single lamellae (figs. 76, 77, 79) 30

29(28)   Gills on abdominal segments 1–2 oval, posterior (ventral) lamella smaller than anterior (dorsal) lamella (fig. 74); claws of middle and hind legs distinctly longer than those of forelegs (fig. 73); rare; mid-boreal, California . . . .*Edmundsius* (p. 133)

Gills on abdominal segments 1–2 subtriangular, broadest near apex, posterior (ventral) lamella subequal to anterior (dorsal) lamella (fig. 75); claws of middle and hind legs slightly longer than those of forelegs (fig. 72); mid-boreal, south to Georgia, Illinois, Arkansas, New Mexico, and California . . . .
. . . .*Siphlonurus* (p. 138)

30(28)   Gills obovate with a sclerotized band (sb) along lateral margin and usually with a similar sclerotized band on or near mesal margin (mb) (figs. 76, 77); maxillae with crown of pectinate

---

NYMPHAL CHARACTERS. Fig. 78. Head and enlarged pectinate spine,*Ameletus oregonensis.* Fig. 79. Gill 4, *Siphlonisca aerodromia.* Fig. 80. Abdomen, segments 4–6, *Baetodes* sp. Fig. 81. Abdomen, segments 4–6, *Baetodes* sp. Fig. 82. Claw, *Dactylobaetis warreni.* Fig. 83. Tarsus and claw, *Dactylobaetis* sp. Fig. 84. (a) Caudal filaments, (b) detail of section, *Centroptilum* sp. Fig. 85. Head (ventral), *Centroptilum* sp. Fig. 86. Labial palpus, *Centroptilum* sp. Fig. 87. Gill 1, *Centroptilum* sp. Fig. 88. Gill 1, *Cloeon* sp. Fig. 89. Gill 1, *Cloeon dipterum.* Fig. 90. Gill 4, *Centroptilum* sp. Fig. 91. Gills 4 on abdomen, *Cloeon triangulifer.* Fig. 92. Gills 4 on abdomen, *Centroptilum* sp. Fig. 93. Gill 2, *Callibaetis* sp. Fig. 94. Gill 4, *Callibaetis coloradensis.* Fig. 95. Gill 4, *Baetis tricaudatus.* (Fig. 94 from Edmunds, Allen, and Peters, 1963.)

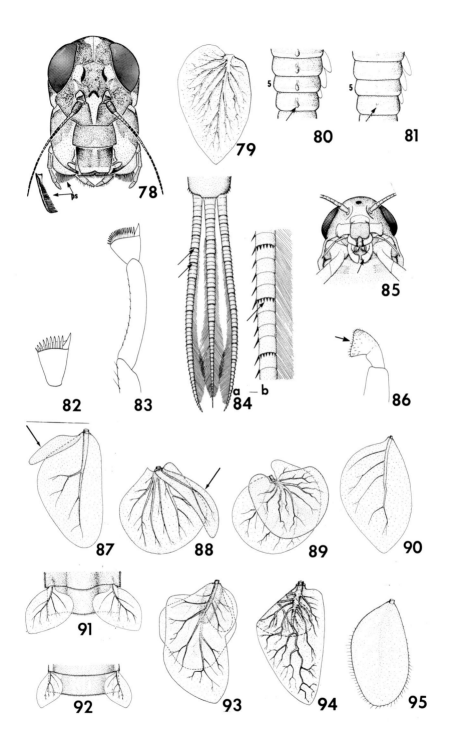

spines (ps) (fig. 78); mid-boreal, south to Georgia, Illinois, New Mexico, and California . . . .*Ameletus* (p. 131)

Gills cordate or subcordate (fig. 79); maxillae without pectinate spines 31

31(30)   Abdominal segments 5–9 greatly expanded laterally; sternum of mesothorax and metathorax each with a median tubercle; rare; mid-boreal, New York . . . .*Siphlonisca* (p. 136)

Abdominal segments 5–9 not greatly expanded; sterna of thorax without tubercles; nymph shown (fig. 391); upper boreal, Canada and western North America south to Wyoming and Utah . . . .*Parameletus* (p. 134)

32(25)   Claws distinctly spatulate with large apical denticles, tarsi distinctly bowed (figs. 82, 83); mid-austral, Central America north to Oregon, Saskatchewan, and Oklahoma . . . .
. . . .*Dactylobaetis* (p. 174)

Claws sharply pointed; denticles, if present, smaller and ventral (figs. 102–110) 33

33(32)   Abdominal gills present on segments 1–5 only, extended ventrally from the pleura; caudal filaments bare or with only a few setae; distinct tubercle or tuft of setae present on each of abdominal terga 1–7, 1–8, or 1–9 (figs. 80, 81); mid-austral, Central America north to Texas, Oklahoma, and Arizona . . . .
. . . .*Baetodes* (p. 164)

Abdominal gills present on segments 1–7 or 2–7, held laterally or somewhat dorsally; caudal filaments usually with fringe of hairs on inner margin; tubercles or tufts of setae absent from abdomen (fig. 64) 34

34(33)   Caudal filaments with a narrow dark band at apex of every third

---

NYMPHAL CHARACTERS. Fig. 96. Labial palpus, *Baetis tricaudatus*. Fig. 97. Labial palpus, *Baetis* sp. (*lapponica* group). Fig. 98. Labial palpus, *Baetis spinosus*. Fig. 99. Labial palpus, *Baetis* sp. Fig. 100. Labial palpus, *Paracloeodes abditus*. Fig. 101. Labial palpus, *Apobaetis indeprensus*. Fig. 102. Tarsus and claw, *Centroptilum* sp. Fig. 103. Tarsus and claw, *Centroptilum* sp. Fig. 104. Tarsus and claw, *Cloeon* sp. Fig. 105. Tarsus and claw, *Cloeon dipterum*. Fig. 106. Tarsus and claw, *Cloeon* sp. Fig. 107. Tarsus and claw, *Paracloeodes abditus*. Fig. 108 Tarsus and claw, *Apobaetis indeprensus*. Fig. 109. Tarsus and claw, *Baetis tricaudatus*. Fig. 110. Tarsus and claw, *Baetis* sp. Fig. 111. Head, *Pseudocloeon* sp. Fig. 112. Head, *Baetis* sp. Fig. 113. Thorax and first abdominal segments, showing hind wing pad beneath forewing pad, *Baetis* sp.

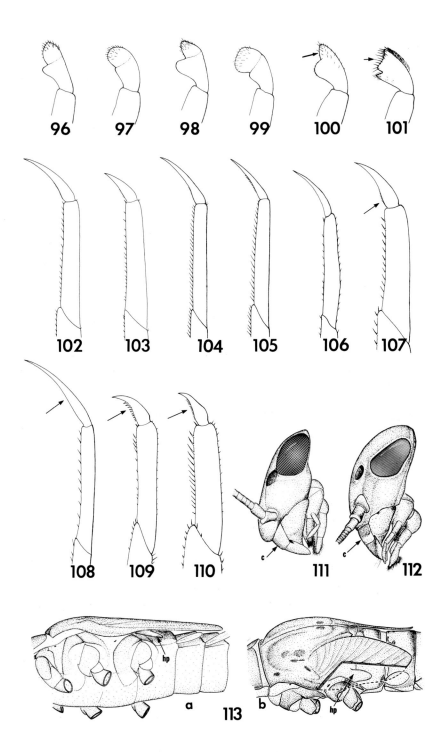

to fifth segment in basal two-thirds (fig. 84); apex of labial palpi simple and truncate (figs. 85, 86); gills usually rather broad, may be simple or with smaller or subequal anterior (dorsal) portion on some segments; trachea of gills, if conspicuous, usually palmate or with branches better developed on mesal side (figs. 87–92). [See discussion of taxonomy under *Cloeon*.] 35

Caudal filaments without narrow dark bands arranged as above; apex of labial palpi variable, not simple and truncate (figs. 96–101); gills on abdominal segments 4–7 longer than broad, simple, or with smaller posterior (ventral) portion; trachea of gills, if conspicuous, pinnate and usually almost symmetrical (figs. 95, 93, 94) 39

35(34)   Gills on abdominal segments 1–2 appear double, with anterior (dorsal) portion only slightly smaller than posterior (ventral) portion (fig. 89); mid-boreal, south to Idaho, Colorado, Texas, and Florida . . . .*Cloeon*, in part (p. 172)

Gills on abdominal segments 1–2 simple or with anterior (dorsal) portion forming a flap smaller than posterior (ventral) portion (figs. 91, 92, 87, 88) 36

36(35)   Gills on abdominal segments 1–2 (sometimes as far posterior as segment 6) with anterior (dorsal) portion forming a small flap (figs. 87, 88) 37

Gills on all abdominal segments simple, without anterior (dorsal) flaps (figs. 91, 92) 38

37(36)   Hind wing pads absent; mid-boreal, south to Idaho, Colorado, Texas, and Florida . . . .*Cloeon*, in part (p. 172)

Hind wing pads (hp) present (as in fig. 113a, b); mid-boreal, south to northern Mexico . . . . . .*Centroptilum*, in part (p. 169)

38(36)   Hind wing pads absent; gills on abdominal segment 4 almost as broad as long (fig. 91); mid-boreal, south to Idaho, Colorado, Texas, and Florida . . . .*Cloeon*, in part (p. 172)

Hind wing pads (hp) present (as in fig. 113a, b); gills on abdominal segment 4 much longer than broad (fig. 92); nymph shown (fig. 400); mid-boreal, south to northern Mexico . . . .*Centroptilum*, in part (p. 169)

39(34)   Gills on one or more abdominal segments with recurved posterior (ventral) flap smaller than anterior (dorsal) portion; flaps may be almost as large as anterior (dorsal) portion and may be re-folded so that gills on anterior abdominal segments appear double or triple (figs. 93, 94); nymph shown (fig. 399); upper austral, widespread north to Alaska . . . .*Callibaetis* (p. 165)

Gills on abdominal segments 1–7 or 2–7 consist of flat, simple lamellae without recurved flaps (fig. 95). 40

40(39)   Terminal filament one-eighth as long as cerci or longer, usually one-fourth as long as cerci to fully as long as cerci 41

Terminal filament apparently absent or less than one-eighth as long as cerci 44

41(40)   Terminal filament shorter and thinner than cerci; widespread distribution. [See discussion of taxonomy under Baetidae.] . . . . . .
. . . .*Baetis*, in part (p. 158)

Terminal filament subequal to cerci in length and thickness . . .
. 42

42(41)   Claws long and slender, about half or more than half as long as respective tarsi, either without denticles or with denticles minute (figs. 107, 108); hind wing pads absent 43

Claws short and often stout, clearly less than half as long as respective tarsi, denticles usually rather short, stout, and progressively longer toward tip of claw (figs. 109, 110); hind wing pads (hp) present (fig. 113a, b); widespread distribution .
. . . .*Baetis*, in part (p. 158)

43(42)   Claws almost as long as respective tarsi (fig. 108); apex of labial palpi truncate (fig. 101); rare; mid-austral (?), California only
. . . .*Apobaetis* (p. 157)

Claws about half as long as respective tarsi (fig. 107); apex of labial palpi conical (fig. 100); mid-austral, north to California, Wyoming, Minnesota, Mississippi, and Georgia . . . .
. . . .*Paracloeodes* (p. 177)

44(40)   Fore coxae each with a single filamentous gill; mid-boreal, eastern North America, Quebec, and Ontario south to Georgia . . .
. . . .*Heterocloeon* (p. 176)

Fore coxae without filamentous gills 45

45(44)   Nymph distinctly depressed; mouthparts relatively short and clypeus (c) deflexed under head; in lateral view, antennae appear to be inserted in lower one-third of head (fig. 111); sternum of abdominal segment 2 only one-fourth as long as broad; nymph shown (fig. 401); mid-boreal, south to Florida, Tennessee, Illinois, and Utah. [See discussion of taxonomy under Baetidae.]
.... *Pseudocloeon*, in part (p. 178)

Nymph not as depressed; mouthparts longer and clypeus (c) in similar plane with frons; in lateral view, antennae appear to be inserted about midway up head (fig. 112); sternum of abdominal segment 2 more than one-third as long as broad. [See discussion of taxonomy under Baetidae.] 46

46(45)   Developing hind wing pads (hp) present although sometimes minute (fig. 113a, b); widespread distribution ....
.... *Baetis*, in part (p. 158)

Developing hind wing pads absent; mid-boreal, south to Florida, Tennessee, Illinois, and Utah ....
.... *Pseudocloeon*, in part (p. 178)

47(11)   Two caudal filaments present; lamellate portion of abdominal gills 2–7 oval, fibrilliform portion well developed (fig. 115); nymph shown (fig. 402); mid-austral, Central America north to Utah, Wyoming, and Saskatchewan .... *Lachlania* (p. 183)

Three caudal filaments present; lamellate portion of abdominal gills 2–7 lanceolate, fibrilliform portion absent (fig. 114); mid-austral, Central America north to South Carolina, Indiana, Nebraska, and Utah .... *Homoeoneuria* (p. 182)

48(20)   Two well-developed caudal filaments present, terminal filament rudimentary or absent 49

Three well-developed caudal filaments present 51

49(48)   Well-developed paired tubercles present on abdominal terga 1–9 (fig. 116); upper boreal, northwestern North America south to California, Nevada, and Montana .... *Ironodes* (p. 197)

Abdominal terga without paired tubercles (fig. 117); mid-boreal, widespread in North America south to Honduras . .*Epeorus*, 50

50(49)   Abdominal terga with dense median row of setae (fig. 117); gills and body usually purplish-brown; mid-boreal, northwestern

North America south to California, Nevada, Wyoming, and Vera Cruz (Mexico) . . . . .subgenus *Ironopsis* (p. 193)

Abdominal terga with median row of setae absent or poorly defined; gills and body usually pale to medium brown; nymph shown (fig. 403); mid-boreal, widespread in North America south to Costa Rica . . . . .subgenus *Iron* (p. 192)

51(48)   Gills on abdominal segments 1 (g_1) and 7 (g_7) enlarged and meet or almost meet beneath abdomen to form ventral disc (fig. 118); mid-boreal, widespread in North America south to Guatemala . . . .*Rhithrogena* (p. 198)

Gills on abdominal segments 1 and 7 do not meet beneath abdomen, usually smaller than intermediate pairs 52

52(51)   Gills on abdominal segment 7 reduced to slender filaments; trachea, if present in gill 7, with few or no lateral branches (fig. 119). [Note: One species of *Heptagenia* has minute gills on abdominal segment 7 (fig. 121), much smaller than gill 7 of *Stenonema* and *Stenacron*.] 53

Gills on abdominal segment 7 similar to preceding pairs but smaller; trachea of gill 7 with lateral branches (fig. 120) . . .54

53(52)   Gills on abdominal segments 1–6 with apex pointed (fig. 122); maxillae with stout spines on crown of galea-lacinia (fig. 123); mid-boreal, eastern North America south to Florida, west to Arkansas and Minnesota . . . .*Stenacron* (p. 200)

Gills on abdominal segments 1–6 with apex rounded or truncate (figs. 126, 127); maxillae with setae or plumose hairs on crown of galea-lacinia (figs. 124, 125); lower boreal, widespread in North America south to Panama . . . .*Stenonema* (p. 202)

54(52)   Front of head distinctly emarginate medially, maxillary palpi (mp) normally partially visible at sides of head from dorsal view (fig. 128); fibrilliform portion (f) of abdominal gills absent or reduced to few tiny filaments (fig. 129); mid-boreal, western North America south to California and New Mexico, eastern North America south to Georgia . . . .*Cinygmula* (p. 188)

Front of head entire or feebly emarginate, not as above; maxillary palpi normally not visible from dorsal view; fibrilliform portion (f) of abdominal gills distinct on segments 1–6, may be absent on 7 (figs. 117, 120, 121) 55

55(54)   Gill lamellae (lm) on abdominal segment 1 distinctly smaller than those on segment 2; fibrilliform portion (f) of gill 1 longer than lamella (lm) (fig. 131); labrum narrow, extending not more than one-fourth of distance along anterior margin of head; upper boreal, northwestern North America south to California, Nevada, Idaho, and Wyoming . . . .*Cinygma* (p. 186)

Gill lamellae (lm) on abdominal segment 1 only slightly shorter than those on segment 2; fibrilliform portion (f) of gill 1 usually subequal to or shorter than lamella (lm) (fig. 132); labrum broad, extending two-thirds to three-fourths of distance along anterior margin of head; lower boreal, widespread south to Guatemala . . . .*Heptagenia* (p. 194)

56(23)   Gills on abdominal segments 1–3 single, without recurved flap (fig. 130); abdomen slender, tergum 9 about 1¼ times as wide as long; upper boreal, Canada south to Michigan and Maine . . . .*Metretopus* (p. 148)

Gills on abdominal segments 1–2 or 1–3 with posterior (ventral) recurved flap or flaps (figs. 133, 134); abdomen more robust, tergum 9 almost twice as wide as long; nymph shown (fig. 395); mid-boreal, Canada south to Illinois and Florida . . . .*Siphloplecton* (p. 149)

57(17, 24)   Labrum (lb) as broad as or broader than width of head capsule (fig. 135) 58

Labrum (lb) narrower than width of head capsule (figs. 136–138) 59

58(57)   Abdominal gills with margins fringed (fig. 139); mid-austral, north to Washington, Alberta, Saskatchewan, North Dakota, and Indiana . . . .*Traverella* (p. 237)

Abdominal gills not fringed, each lamella terminating in single acute median filament (fig. 140); lower austral, north to Honduras . . . .*Hermanella* (p. 221)

NYMPHAL CHARACTERS. Fig. 114. Gill 4, *Homoeoneuria* sp. Fig. 115. Gill 4, *Lachlania powelli*. Fig. 116. Apex of abdomen, *Ironodes* sp. Fig. 117. Abdomen, *Epeorus (Ironopsis) grandis*. Fig. 118. Abdomen (ventral), *Rhithrogena robusta*. Fig. 119. Apex of abdomen, *Stenacron interpunctatum*. Fig. 120. Apex of abdomen, *Heptagenia solitaria*. Fig. 121. Apex of abdomen, *Heptagenia* sp. Fig. 122. Gill 4, *Stenacron* sp. Fig. 123. Maxilla, *Stenacron* sp. Fig. 124. Maxilla, *Stenonema tripunctatum*. Fig. 125. Maxilla, *Stenonema* sp. Fig. 126. Gill 4, *Stenonema tripunctatum*. Fig. 127. Gill 4, *Stenonema pulchellum*. (Figs. 122, 123 from Jensen, 1974.)

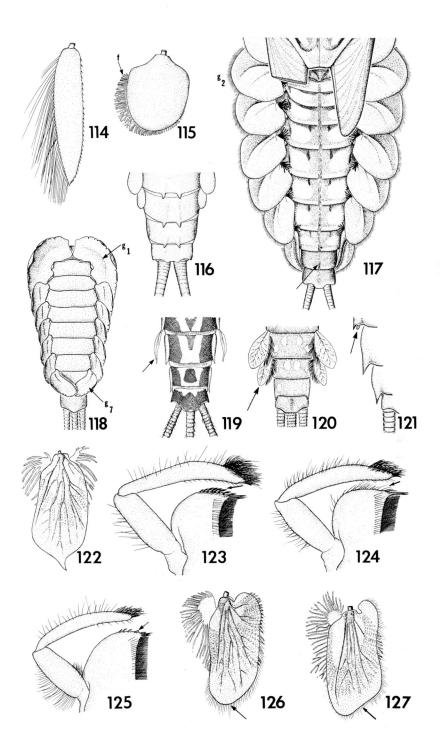

59(57)   Abdominal gills bilamellate with margins fringed (figs. 141, 142); lower austral, north to Costa Rica . . . .*Ulmeritus* and *Ulmeritus* ally (p. 238)

Abdominal gills variously shaped, not bilamellate with margins fringed (figs. 143–154, 157–159) 60

60(59)   Abdominal gills 2–7 consist of clusters of slender filaments (fig. 143); mid-boreal, eastern North America south to Florida . . . .*Habrophlebia* (p. 218)

Abdominal gills 2–7 forked or bilamellate, not as above (figs. 144b, 145, 146b, 147b, 148–150, 151b, 152b, 153, 154, 157–159) 61

61(60)   Gills on abdominal segment 1 different in structure from those on succeeding segments, those on segment 1 either forked or unforked linear lamellae (figs. 144a, 146a, 147a), those on succeeding segments broadly bilamellate with three-lobed apexes (figs. 144b, 145, 146b, 147b) 62

Gills on abdominal segment 1 similar in structure to those on succeeding segments, sometimes smaller or more slender (figs. 151, 152) 64

62(61)   Gills on abdominal segment 1 unforked (fig. 144a); gills on segments 2–7 with anterior (dorsal) lamella terminating in single broadened filament or in three slender filaments (figs. 144b, 145); nymph shown (fig. 409); lower boreal, widespread south to Panama . . . .*Choroterpes*, in part (p. 214)

Gills on abdominal segment 1 forked (figs. 146a, 147a); gills on segments 2–7 variable (figs. 146b, 147b) 63

63(62)   Gills on abdominal segment 1 almost symmetrically forked (fig. 147a); gills on segments 2–7 terminated in single slender filament flanked by two blunt lobes (fig. 147b); nymph shown (fig. 411); mid-boreal, south to Oregon, Utah, Illinois, Ohio, and Florida . . . .*Leptophlebia* (p. 226)

NYMPHAL CHARACTERS. Fig. 128. Head, *Cinygmula* sp. Fig. 129. Gill 4, *Cinygmula* sp. Fig. 130. Gill 2, *Metretopus borealis*. Fig. 131. Gills 1 and 2, *Cinygma integrum*. Fig. 132. Gills 1 and 2, *Heptagenia criddlei*. Fig. 133. Gill 2, *Siphloplecton interlineatum*. Fig. 134. Gill 2, *Siphloplecton speciosum*. Fig. 135. Head, *Traverella albertana*. Fig. 136. Head, *Hagenulopsis* sp. Fig. 137. Head, *Homothraulus* sp. Fig. 138. Head, *Leptophlebia* sp. Fig. 139. Gill 4, *Traverella albertana*. Fig. 140. Gill 4, *Hermanella* sp. Fig. 141. Gill 4, *Ulmeritus* ally. (Fig. 139 from Edmunds, Allen, and Peters, 1963.)

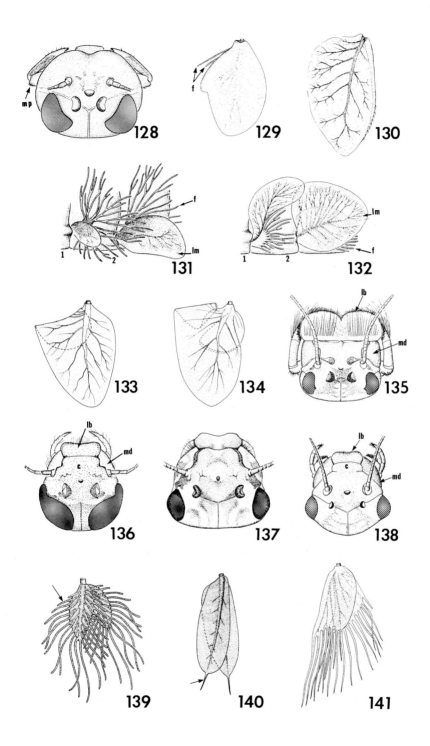

Gills on abdominal segment 1 asymmetrically forked (fig. 146a); gills on segments 2–7 terminated in three filaments, the middle one longest and broadest (fig. 146b); lower boreal, northern Mexico and Texas . . . .*Choroterpes*, in part (p. 214)

64(61)   Gills on middle abdominal segments shaped as in figs. 148, 157; lateral tracheal branches conspicuous 65

Gills on middle abdominal segments not as above, either narrower with lateral tracheal branches inconspicuous, or more deeply cleft (figs. 149, 150, 151b, 152b) 66

65(64)   Labrum with moderately deep V-shaped median emargination (fig. 155); small row of spinules present on posterior margins of abdominal terga 6–10 or 7–10 only (fig. 157); nymph shown (fig. 410); mid-boreal, northeastern North America south to Florida, Illinois, and Oklahoma . . . .*Habrophlebiodes* (p. 218)

Labrum with shallow, variably shaped median emargination (figs. 156, 161); small row of spinules present on posterior margins of abdominal terga 1–10; mid-boreal, south to Florida, Texas, New Mexico, and California . . . .
. . . .*Paraleptophlebia*, in part (p. 230)

66(64)   Labrum about two-thirds as wide as head, distinctly wider than clypeus; gills variable, but each lamella up to one-fourth as broad as long (figs. 158, 159); posterolateral spines present on abdominal segments 2–9, usually small on segments 2–3 or 2–4; nymph shown (fig. 413); mid-austral, north to Arizona, New Mexico, and Texas . . . .*Thraulodes* (p. 234)

Labrum less than half as wide as head, about as wide as clypeus (c) (similar to figs. 136–138); each gill lamella usually no more than one-eighth as broad as long (figs. 149–154); posterolateral spines usually present on abdominal segments 6–9 or 8–9 or on segment 9 only. [For species from Texas to Central America, see discussion of taxonomy under *Homothraulus*, p. 224.] 67

NYMPHAL CHARACTERS. Fig. 142. Gill 4, *Ulmeritus*. Fig. 143. Gill 5, *Habrophlebia vibrans*. Fig. 144. (a) Gill 1, (b) gill 4, *Choroterpes albiannulata*. Fig. 145. Gill 4, *Choroterpes kossi?* Fig. 146. (a) Gill 1, (b) gill 4, *Choroterpes mexicanus?* Fig. 147. (a) Gill 1, (b) gill 4, *Leptophlebia gravastella*. Fig. 148. Gill 4, *Habrophlebiodes americana*. Fig. 149. Gill 4, *Paraleptophlebia heteronea*. Fig. 150. Gill 4, *Paraleptophlebia* sp. Fig. 151. (a) Gill 1, (b) gill 4, *Paraleptophlebia debilis*. Fig. 152. (a) Gill 1, (b) gill 4, *Paraleptophlebia* sp. (Texas). Fig. 153. Gill 4, *Hagenulopsis* sp. Fig. 154. Gill 4, *Homothraulus* sp. Fig. 155. Labrum, *Habrophlebiodes americana*. Fig. 156. Labrum, *Paraleptophlebia* sp. Fig. 157. Abdomen, *Habrophlebiodes brunneipennis*. Fig. 158. Gill 4, *Thraulodes* sp. Fig. 159. Gill 4, *Thraulodes* sp. (Mexico).

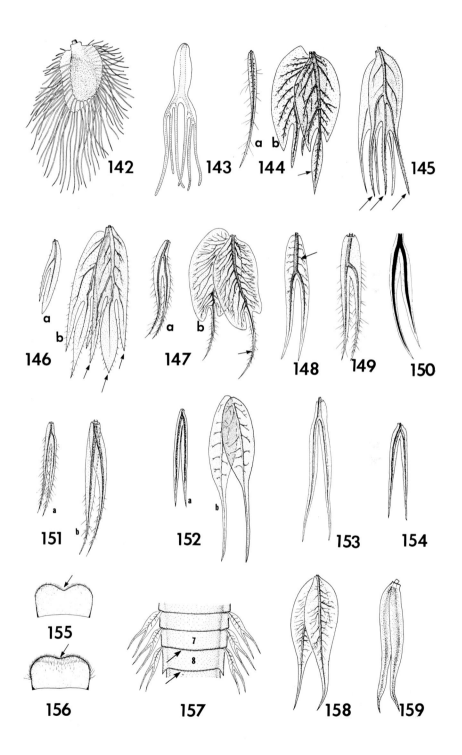

142 143 144 145

146 147 148 149 150

151 152 153 154

155

156 157 158 159

67(66)   Distribution in Central America; labrum with denticles on median emargination (fig. 160); claws with prominent subapical
denticle and smaller denticles basally (figs. 163–165); lower austral, north to Honduras 68

Distribution in United States and Canada; labrum without denticles on median emargination (figs. 156, 161); claws with slender
denticles subequal in size (fig. 162); mid-boreal, south to
Florida, Texas, New Mexico, and California . . . .
. . . .*Paraleptophlebia*, in part (p. 230)

68(67)   Developing hind wing pads present; posterolateral projections on
abdominal segments 8–9; lower austral, north to Darien (Pana-
ma) . . . .*Hermanellopsis* (p. 222)

Developing hind wing pads absent; posterolateral projections on
abdominal segments 6–9; lower austral, north to Honduras . . .
. . . .*Hagenulopsis* (p. 221)

69(15)   Lamellate gills present on abdominal segments 3–7 (figs. 50,
167) 70

Lamellate gills present on abdominal segments 4–7 (fig.
183) 76

70(69)   Cerci one-fourth to three-fourths as long as terminal filament;
upper boreal, northwestern North America south to California,
Idaho, and Wyoming . . . . .subgenus *Caudatella* (p. 243)

Cerci and terminal filament subequal in length 71

71(70)   Distinct tubercles present on ventral (leading) edge of fore femora (fig. 171); claws with only one to four denticles; mid-boreal,
western North America south to Baja California and New
Mexico, eastern North America south to Michigan and Georgia
. . . . .subgenus *Drunella*, in part (p. 244)

NYMPHAL CHARACTERS. Fig. 160. (a) Labrum, (b) detail of median emargination, *Hagenulopsis*.
Fig. 161. Labrum, *Paraleptophlebia* sp. Fig. 162. Claw, *Paraleptophlebia* sp. Fig. 163. Claw,
Hermanellopsis sp. Fig. 164. Claw, *Hagenulopsis* species no. 1. Fig. 165. Claw, *Hagenulopsis*
species no. 2. Fig. 166. Claw, *Homothraulus* sp. Fig. 167. Nymph (legs removed), *Ephemerella
(Drunella) grandis*. Fig. 168. Maxilla, *Ephemerella (Serratella) tibialis*. Fig. 169. Maxilla,
Ephemerella (Serratella) teresa. Fig. 170. Maxilla, *Ephemerella (Ephemerella) inermis*. Fig.
171. Foreleg, *Ephemerella (Drunella) doddsi*. Fig. 172. Foreleg, *Ephemerella (Serratella) ser-
ratoides*. Fig. 173. Foreleg, *Ephemerella (Ephemerella) septentrionalis*. Fig. 174. Caudal filaments, *Ephemerella (Ephemerella) rotunda*. Fig. 175. Caudal filaments, *Ephemerella (Serratella) tibialis*. (Figs. 172, 175 from Allen and Edmunds, 1963a; figs. 173, 174 from Allen and
Edmunds, 1965.)

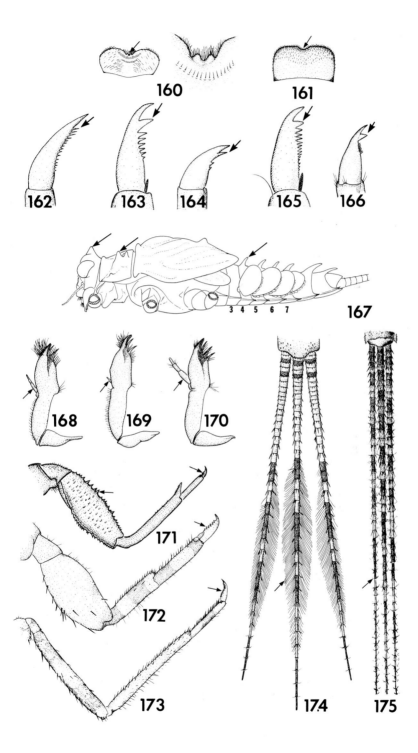

160

161

162 163 164 165 166

167

3 4 5 6 7

168 169 170

171

172

173

174 175

Tubercles absent on ventral (leading) edge of fore femora (figs. 172, 173, 176, 177); claws usually with more than four denticles 72

72(71)   Abdominal sterna 3–8 with adhesive disc of long hair; mid-boreal, Washington, Oregon, and California . . . . .subgenus *Drunella*, in part (p. 244)

Abdominal sterna may have long hair, but only sparse, not forming adhesive disc 73

73(72)   Head, thorax, *and* abdomen with well-developed tubercles (fig. 167); mid-boreal, western North America south to California, Arizona, and New Mexico . . . . .subgenus *Drunella*, in part (p. 244)

Tubercles smaller than above or absent, only rarely present on head, thorax, and abdomen 74

74(73)   Legs long and thin (fig. 173); abdominal terga 2–7 each with small median tubercle; mid-boreal, eastern North America south to North Carolina and Tennessee . . . . .subgenus *Ephemerella*, sensu stricto, in part (p. 247)

Legs shorter and more robust (fig. 172); abdominal terga either without tubercles or with paired submedian tubercles 75

75(74)   Caudal filaments with whorls of short spines at apex of each segment and with only sparse intersegmental setae or none (fig. 175); maxillary palpi absent or reduced in size (figs. 168, 169); mid-boreal, western North America south to Baja California and New Mexico, eastern North America south to Florida, Mississippi, and Arkansas . . . . .subgenus *Serratella* (p. 250)

Caudal filaments with or without whorls of spines at apex of each segment but with numerous intersegmental setae (fig. 174); maxillary palpi well developed (fig. 170); nymph shown (fig. 415); mid-boreal, widespread south to California, New Mexico,

NYMPHAL CHARACTERS. Fig. 176. Middle leg, *Ephemerella (Timpanoga) hecuba*. Fig. 177. Middle leg, *Ephemerella (Eurylophella) bicolor*. Fig. 178. Foreleg, *Tricorythodes minutus*. Fig. 179. (a) Foreleg, (b) spine (enlarged), *Leptohyphes apache*. Fig. 180. (a) Foreleg, (b) spine (enlarged), *Leptohyphes robacki*. Fig. 181. Gill, segment 2, *Tricorythodes minutus*. Fig. 182. Gill, segment 2, *Tricorythodes edmundsi*. Fig. 183. Abdomen, *Ephemerella (Dannella) simplex*. Figs. 184–186. Gills, segment 2, *Leptohyphes* spp. (Peru). Fig. 187. Head, *Caenis* sp. (Fig. 177 from Allen and Edmunds, 1963b; figs. 179, 180 from Allen, 1967; fig. 183 from Allen and Edmunds, 1962a; figs. 184–186 from Allen and Roback, 1969.)

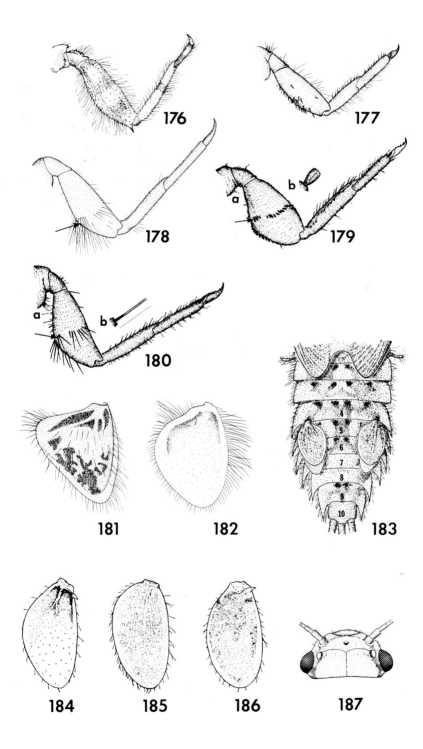

Oklahoma, Alabama, and Florida . . . .
. . . . .subgenus *Ephemerella*, sensu stricto, in part (p. 247)

76(69)   Apex of each femur with acute point (fig. 176); head widest near
anterior margin; nymph (fig. 417); mid-boreal, western North
America south to California and New Mexico . . . .
. . . . .subgenus *Timpanoga* (p. 251)

Apex of each femur rounded or angulate, not with acute point
(fig. 177); head widest at eyes 77

77(76)   Abdominal terga without paired submedian tubercles (fig. 183);
mid-boreal, eastern North America south to Illinois and Florida
. . . . .subgenus *Dannella* (p. 244)

Abdominal terga with paired submedian tubercles, at least on
segments 3–7 78

78(77)   Abdominal segments 5, 6, and 7 conspicuously shortened, segment 9 conspicuously elongated so that length of segment 9 is
about equal to combined lengths of segments 5, 6, and 7; nymph
shown (fig. 416); mid-boreal, western North America south to
California, eastern North America south to Missouri and
Florida . . . . .subgenus *Eurylophella* (p. 248)

Abdominal segments 2–9 all about equal length; segment 9 may
be slightly elongate but shorter than combined lengths of segments 5 and 6 only; nymph shown (fig. 414); mid-boreal, western North America south to California and New Mexico, eastern
North America south to Florida . . .subgenus *Attenella* (p. 241)

79(13)   Femora with long setae (fig. 178); operculate abdominal gill 2
triangular or subtriangular in shape, over two-thirds as wide as
long (figs. 181, 182); nymph shown (fig. 419); upper austral,
north to British Columbia, Saskatchewan, Quebec, and Newfoundland . . . .*Tricorythodes* (p. 256)

Femora with spines (figs. 179, 180); operculate abdominal gill 2
oval in shape, less than two-thirds as wide as long (figs. 184–186);
nymph shown (fig. 418); mid-austral, north to Utah, Texas, and
Maryland . . . .*Leptohyphes* (p. 254)

NYMPHAL CHARACTERS. Fig. 188. Foreleg, *Euthyplocia* sp. Fig. 189. Foreleg, *Campylocia* sp.
Fig. 190. Head, *Ephemera simulans*. Fig. 191. Gill 1, *Hexagenia limbata*. Fig. 192. Head,
Hexagenia limbata. Fig. 193. Gill 1, *Litobrancha recurvata*. Fig. 194. Head, *Litobrancha
recurvata*.

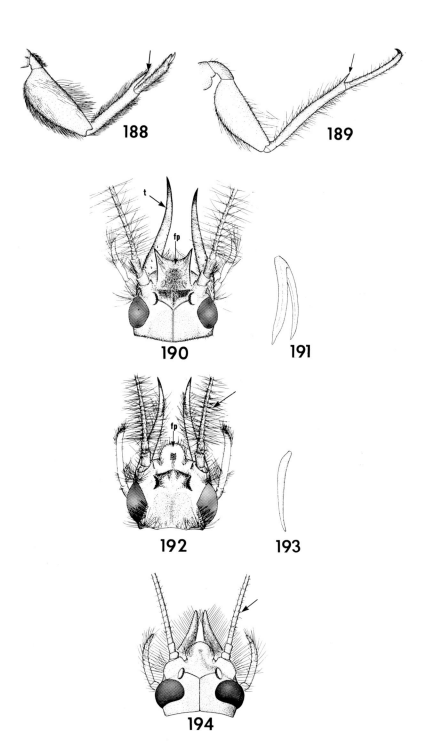

188

189

190

191

192

193

194

80(14)   Head with three prominent ocellar tubercles; maxillary and labial palpi two-segmented; nymph shown (fig. 421); lower boreal, widespread in eastern and central North America west to Idaho and Utah, south to Panama . . . .*Brachycercus* (p. 264)

Head without ocellar tubercles (fig. 187); maxillary and labial palpi three-segmented; lower boreal, widespread south to Panama . . . .*Caenis* (p. 266)

81(8)   Apical extension of tibiae of forelegs nearly half length of tarsi, tarsi extending beyond bases of claws (fig. 188); antennae three times longer than mandibular tusks; lower austral, Central America north to Vera Cruz (Mexico) . . .*Euthyplocia* (p. 279)

Apical extension of tibiae of forelegs one-fourth length of tarsi, tarsi with claws at apex (fig. 189); antennae shorter, rarely longer than mandibular tusks; nymph shown (fig. 426); lower austral, Central America . . . .*Campylocia* (p. 279)

82(6)   Frontal process (fp) of head bifid (fig. 190); mid-boreal, south to Utah, Colorado, Oklahoma, and Florida . . .*Ephemera* (p. 282)

Frontal process (fp) of head entire, either truncate, rounded, or conical (figs. 192, 194) 83

83(82)   Gills on abdominal segment 1 forked (fig. 191); antennae with whorls of long setae (fig. 192); upper austral, Panama north to British Columbia, Northwest Territories, Manitoba, Ontario, and New Brunswick . . . .*Hexagenia*, 84

Gills on abdominal segment 1 single (fig. 193); antennae with short scattered setae (fig. 194); mid-boreal, eastern North America from Ontario and Quebec south to North Carolina . . . .*Litobrancha* (p. 291)

84(83)   Distribution upper austral: Jalisco (Mexico) north to British Columbia, Northwest Territories, Manitoba, Ontario, and New Brunswick . . . . .subgenus *Hexagenia* (p. 288)

Distribution lower austral: Central America north to southern Mexico . . . . .subgenus *Pseudeatonica* (p. 290)

85(7)   Mandibular tusks with single prominent subapical tubercle on median margin although another tubercle may also occur basal to this (fig. 33); mid-austral, Central America north to Nebraska, Manitoba, South Carolina, and Florida . . . . . *Tortopus* (p. 300)

Mandibular tusks with a prominent basal or subbasal tubercle on median margin and several to many smaller apical crenations; nymph shown (fig. 432); mid-austral, Central America north to Texas . . . . *Campsurus* (p. 298)

Key to Adults

NOTE: Unless specified otherwise, the venation referred to in the couplets is that of the forewing.

1   Wing venation greatly reduced, apparently only three or four longitudinal veins behind R_1 (figs. 195a, 196a); body black . . . . .Oligoneuriidae, 41

    Wing venation complete or only moderately reduced, numerous longitudinal veins present behind R_1 (figs. 197–201, 217–224); body color variable 2

2(1) Penes of male longer than forceps (fig. 204); antennae of female inserted on prominent anterolateral projections (fig. 202); four or more long cubital intercalaries usually present (fig. 197a); rare; mid-boreal, southeastern United States south to Florida . . . . .Behningiidae, *Dolania* (p. 274)

    Penes of male shorter than forceps (figs. 205–206, 208); antennae of female not inserted as above (fig. 203); three or fewer long cubital intercalaries present (figs. 198a–201a) 3

3(2) Base of veins MP_2 *and* CuA strongly divergent from base of vein MP_1 (figs. 198a–201a, 217a); hind wings with numerous veins and crossveins; vein MA of hind wings unforked (figs. 198b–201b, 217b) 4

    Base of veins MP_2 *and* CuA little divergent from vein MP_1 (vein MP_2 only may diverge from MP_1), fork usually more symmetrical (figs. 218–220, 221a–224a, 298a); hind wings variable, may be reduced or absent; vein MA of hind wings unforked or forked 10

4(3) Costal angulation of hind wings acute or at right angles (fig. 198b); vein A_1 of forewing unforked; costal crossveins basal to bullae weak or atrophied (fig. 198a); rare, except locally in southeastern United States; mid-boreal, southeastern Canada west to Michigan and south to Florida . . . .
. . . . .Neoephemeridae, *Neoephemera* (p. 261)

Costal angulation of hind wings usually rounded (figs. 199b–201b, 217b); if nearly acute or at right angles, forewing vein A_1 forked near margin (fig. 217a); costal crossveins basal to bullae well developed (figs. 199a–201a, 217a) 5

5(4) Middle and hind legs of male and all legs of female feeble, nonfunctional; color usually pale; wings often somewhat translucent and colorless or with gray or purplish gray shading . . . . . .
. . . . .Polymitarcyidae, 6

All legs of both sexes well developed, functional; color variable 7

6(5) Outer margin of wings with a dense network of reticulate marginal veinlets and intercalaries (fig. 199); genital forceps of male four-segmented (fig. 205); mid-boreal, widespread in southern Canada and northern United States south to Georgia, Ohio, Kansas, and New Mexico . . . . .Polymitarcyinae, *Ephoron* (p. 295)

Outer margin of wings with few veinlets or marginal intercalaries (fig. 200); genital forceps of male two-segmented (fig. 206) . .
. . . . .Campsurinae, 85

7(5) Distribution Central America and southern Mexico: Vein MA_2 1 1/3 to 3 times longer than base of MA (fig. 201a); genital forceps of male with one long basal segment and no terminal segment or one short terminal segment (figs. 326, 327); lower austral, north to Vera Cruz (Mexico) . . . . . .Euthyplociidae, 81

Distribution widespread: Vein MA_2 shorter than, subequal to, or only slightly longer than base of MA (figs. 217a, 339a–341a); genital forceps of male either with two long basal segments and one or two short terminal segments, *or* with one long basal segment and two short terminal segments (figs. 208, 214, 346–348) 8

8(7) Vein A_1 forked near wing margin (fig. 217a); genital forceps of male with one long basal segment (fig. 208); abdomen usually yellowish, in some species with reddish lateral stripes or spots on terga; mid-boreal, eastern North America west to Nebraska and Kansas . . . . .Potamanthidae, *Potamanthus* (p. 276)

Vein A_1 unforked, attached to hind margin by three or more veinlets (figs. 339a–341a); genital forceps with two long basal segments (figs. 214, 346–348); abdomen of most species with striking dark pattern on terga and sterna . . . . . .Ephemeridae, 9

9(8)   Pronotum of male well developed, no more than twice as wide as long (figs. 209, 210); penes variable, not long and tubular (figs. 346–348, 386); caudal filaments of female longer than body . . . . .Ephemerinae, 82

Pronotum of male reduced, about three times as wide as long (fig. 211); penes long and tubular (fig. 214); caudal filaments of female shorter than body; mid-austral, central North America, Texas, and Florida north to Manitoba . . . .
. . . . .Pentageniinae, *Pentagenia* (p. 292)

10(3)   Hind wings absent 11
Hind wings present although often minute 14

11(10) Short basally detached single or double marginal intercalaries present in each interspace of wing; veins MA_2 and MP_2 detached basally from their respective stems (figs. 246, 253); two caudal filaments present; penes of male reduced (figs. 359, 361, 364) . . . . .Baetidae, in part, 32

Without short basally detached intercalaries as above; vein MA_2 attached basally, MP_2 attached or detached (figs. 218–220); three caudal filaments present; penes of male well developed (figs. 376, 382, 383, 385) 12

12(11) Wings with several veinlets attaching longitudinal veins to outer margin (fig. 218); eyes of male large, semiturbinate (fig. 212); female with well-developed ovipositor (fig. 213); lower austral, north to Honduras . . . .
. . . . .Leptophlebiidae, in part, *Hagenulopsis* (p. 221)

ADULT CHARACTERS. Fig. 195. (a) Forewing, (b) hind wing, *Lachlania powelli*. Fig. 196. (a) Forewing, (b) hind wing, *Homoeoneuria dolani*. Fig. 197. (a) Forewing, (b) hind wing, *Dolania americana*. Figs. 198. (a) Forewing, (b) hind wing, *Neoephemera purpurea*. Fig. 199. (a) Forewing, (b) hind wing, *Ephoron album*. Fig. 200. (a) Forewing, (b) hind wing, *Campsurus* sp. Fig. 201. (a) Forewing, (b) hind wing, *Campylocia* sp. (Figs. 195, 196 from Edmunds, Berner, and Traver, 1958.)

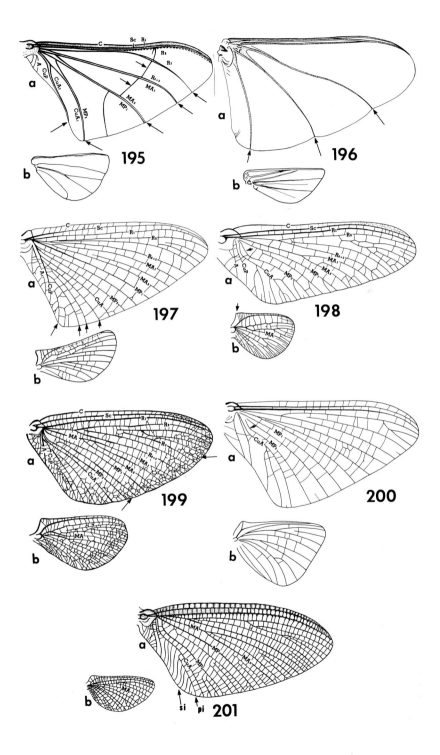

Longitudinal veins of wing not attached to outer margin by vein-lets (figs. 219, 220); eyes of male usually simple, widely separated (fig. 215); female without ovipositor 13

13(12)   Vein MA forming a more or less symmetrical fork; wing veins MP_2 and IMP extend less than three-fourths of distance to base of vein MP_1 (fig. 219); genital forceps of male two-segmented or three-segmented; thorax usually black or gray . . . .
. . . . .Tricorythidae, in part, 78

Vein MA not as above, MA_2 attached basally by a crossvein; wing veins MP_2 and IMP almost as long as vein MP_1 and extend nearly to base (fig. 220); genital forceps of male one-segmented; thorax usually brown . . . . .Caenidae, 80

14(10)   Cubital intercalaries consist of a series of veinlets, often forking or sinuate, attaching vein CuA to hind margin (figs. 221a, 222a); hind tarsi four-segmented . . . . .Siphlonuridae, 15

Cubital intercalaries variable but not as above (figs. 224a, 225a, 230a, 300a, 302a, 330a), sometimes absent (fig. 226a); hind tarsi four-segmented or five-segmented 17

15(14)   Terminal filament present, distinctly longer than tergum 10, may be as long as cerci; hind wings half as long as forewings, strongly triangular in shape and without costal projection (fig. 223); very rare. [See discussion of taxonomy under *Analetris*.] . . . .
. . . . .Acanthametropodinae, *Analetris* (p. 142), and *Acanthametropus* (p. 141)

Terminal filament vestigial, shorter than tergum 10; hind wings less than half as long as forewings, not strongly triangular as above, with some indication of costal projection (figs. 221b, 222b, 228b) 16

16(15)   Vein MP of hind wings forked near margin, stem longer than fork (fig. 222b); forelegs largely or entirely dark, middle and

ADULT CHARACTERS. Fig. 202. Head (female), *Dolania americana*. Fig. 203. Head (female), *Ephoron album*. Fig. 204. Genitalia (male), *Dolania americana*. Fig. 205. Genitalia (male), *Ephoron album*. Fig. 206. Genitalia (male), *Tortopus primus*. Fig. 207. Detail of forceps and appendage, *Tortopus primus*. Fig. 208. Genitalia (male), *Potamanthus*. Fig. 209. Pronotum (male), *Hexagenia limbata*. Fig. 210. Pronotum (male), *Ephemera simulans*. Fig. 211. Pronotum (male), *Pentagenia vittigera*. Fig. 212. Head (male), *Hagenulopsis*. Fig. 213. Apex of abdomen (female), *Hagenulopsis*. Fig. 214. Genitalia (male), *Pentagenia vittigera*. Fig. 215. Head (male) *Caenis*. Fig. 216. Gill remnant at base of foreleg, *Isonychia rufa*. (Fig. 214 from McCafferty, 1972.)

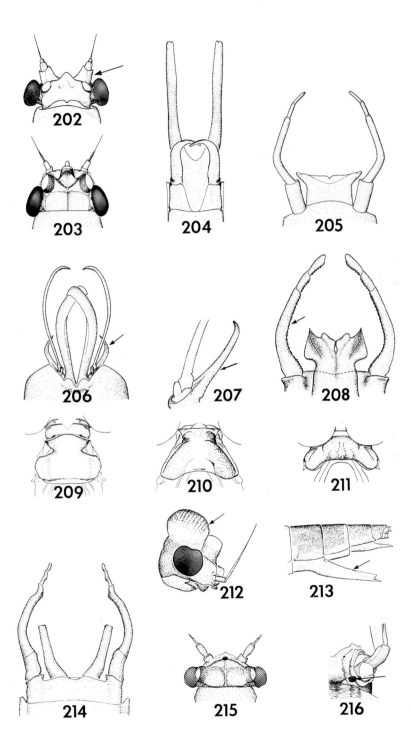

202

203

204

205

206

207

208

209

210

211

212

213

214

215

216

hind legs pale; gill remnants present at bases of fore coxae (fig. 216); lower boreal, southern Canada south to Honduras . . . . . .
. . . . .Isonychiinae, *Isonychia* (p. 144)

Vein MP of hind wings simple and unforked or forked near base, with stem shorter than fork (figs. 229b, 221b); legs variable in color; gill remnants absent . . . . .Siphlonurinae, 28

17(14)   Three well-developed caudal filaments present 18

Two well-developed caudal filaments present, terminal filament rudimentary or absent 21

18(17)   Hind wings small, with two or three simple veins only; costal projection long (1½ to 3 times width of wing) and straight or recurved (figs. 335b, 336b) . . . . .Tricorythidae, in part, 78

Hind wings larger with one or more veins forked; costal projection not as above (figs. 224b, 330b) 19

19(18)   Two pairs of cubital intercalaries present, anterior pair longer; vein A_1 attached to hind margin by a series of veinlets (fig. 224a); mid-boreal, western North America south to Oregon, Utah, and New Mexico. . .Ametropodidae, *Ametropus* (p. 152)

Cubital intercalaries not as above; vein A_1 not attached to hind margin as above (figs. 225a, 299a–302a) 20

20(19)   Short, basally detached marginal intercalaries present between veins along entire outer margin of wings (fig. 225); genital forceps of male with one short terminal segment (figs. 311–316); mid-boreal, widespread in Canada and United States south to northern Sonora and Baja California (Mexico) . . . .
. . . . .Ephemerellidae, *Ephemerella*, 68

No true basally detached marginal intercalaries in positions indicated above, being absent along entire outer margin of wings (figs. 299–302); genital forceps of male with two short terminal segments (figs. 289, 290, 291) . . .Leptophlebiidae, in part, 58

21(17)   Cubital intercalaries absent; vein A_1 terminating in outer margin of wings (fig. 226a); hind wings with numerous long free margi-

---

ADULT CHARACTERS. Fig. 217. (a) Forewing, (b) hind wing, *Potamanthus* sp. Fig. 218. Forewing, *Hagenulopsis* sp. Fig. 219. Forewing, *Tricorythodes minutus*. Fig. 220. Forewing, *Caenis simulans*. Fig. 221. (a) Forewing, (b) hind wing, *Siphlonurus* sp. Fig. 222. (a) Forewing, (b) hind wing, *Isonychia velma*. Fig. 223. (a) Forewing, (b) hind wing, *Analetris eximia*. Fig. 224. (a) Forewing, (b) hind wing, *Ametropus albrighti*. (Fig. 223 from Edmunds and Koss, 1972.)

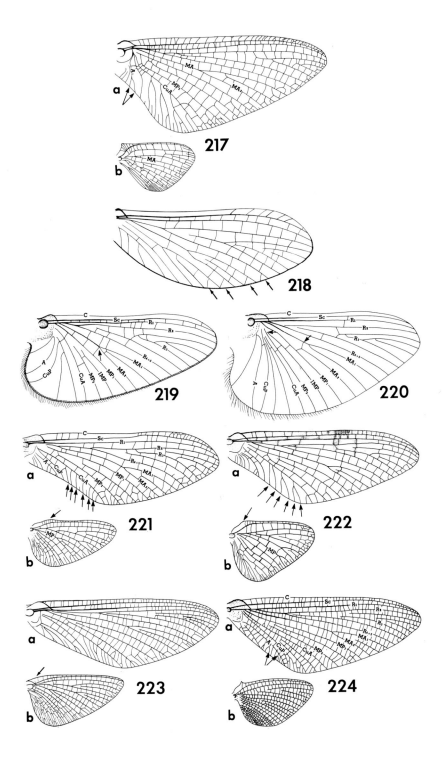

nal intercalaries (fig. 226b); mid-boreal, widely distributed in eastern and central North America, rare in west . . . .
. . . . .Baetiscidae, *Baetisca* (p. 269)
Cubital intercalaries present with vein A_1 terminating in hind margin of wings; hind wings not as above (figs. 247, 275, 297, 298) 22

22(21)   Short basally detached single or double marginal intercalaries present in each interspace of forewings; veins MA_2 and MP_2 detached basally from their respective stems (figs. 230, 231, 247); penes of male reduced (figs. 360, 362, 363); upper portion of eyes of male turbinate (fig. 233) . . . . . .Baetidae, in part, 32

Marginal intercalaries attached basally to other veins; veins MA_2 and MP_2 attached basally (figs. 275a, 297a); penes of male well developed (figs. 279, 284, 372, 373); eyes of male not turbinate 23

23(22)   Hind tarsi apparently four-segmented, basal segment fused or partially fused to tibiae (figs. 234, 236); cubital intercalaries consist of one or two pairs (figs. 297a, 298a) 24

Hind tarsi distinctly five-segmented (fig. 235); cubital intercalaries consist of two pairs (figs. 275a–278a) . . . .
. . . . .Heptageniidae, in part, 25

24(23)   Eyes of male contiguous or nearly contiguous dorsally, fore tarsi three times length of fore tibiae; abdomen of female with apical and basal segments subequal to middle segments in length and width, subanal plate evenly convex . . . . . . .Metretopodidae, 57

Eyes of male separated dorsally by twice width of median ocellus, fore tarsi two times length of fore tibiae; abdomen of female long and slender, apical segments distinctly more elongate and slender than basal segments, subanal plate with median emargination (fig. 237); rare; mid-boreal, central and southeastern North America west to Manitoba, Wyoming and Utah . . . .
. . . .Heptageniidae, in part, Pseudironinae, *Pseudiron* (p. 208)

ADULT CHARACTERS. Fig. 225. (a) Forewing, (b) hind wing, *Ephemerella (Ephemerella) inermis*. Fig. 226. (a) Forewing, (b) hind wing, *Baetisca rogersi*. Fig. 227. (a) Forewing, (b) hind wing, *Arthroplea bipunctata*. Fig. 228. (a) Forewing, (b) hind wing, *Ameletus oregonensis*. Fig. 229. (a) Forewing, (b) hind wing, *Parameletus columbiae*. Fig. 230. (a) Forewing, (b) hind wing, (c) detail of costal projection, *Callibaetis coloradensis*. Fig. 231. (a) Forewing, (b) hind wing, (d) detail of variant, *Centroptilum selanderorum*. Fig. 232. Hind wing, *Centroptilum conturbatum*. (Fig. 226 from Pescador and Peters, 1974.)

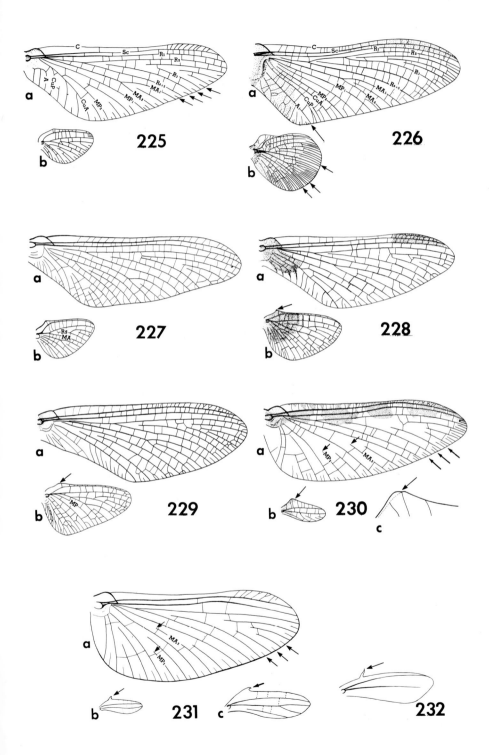

25(23) Vein MA of hind wings simple, unforked, and vein Rs forming a regular fork (fig. 227b); genital forceps of male with three short terminal segments (fig. 238); upper boreal, northeastern North America south to Wisconsin, Ohio, Quebec, and Massachusetts . . . . .Arthropleinae, *Arthroplea* (p. 206)

Vein MA of hind wings usually forked (figs. 272b–278b) [simple, unforked in one western species of *Cinygma* (fig. 271b), but Rs fork is detached basally]; genital forceps of male with two short terminal segments (figs. 279, 283, 284) 26

26(25) Male imagoes: Genital forceps present (figs. 370, 371); eyes large (figs. 265–267) 27

Female imagoes (Anepeorinae females cannot be keyed): Genital forceps absent; eyes smaller than in male . . .Heptageniinae, 50

27(26) Fore tarsi three-fourths or less length of fore tibiae; male genitalia as in fig. 239; rare; mid-boreal, central North America southeast to Georgia and west to Utah . . . .
. . . . .Anepeorinae, *Anepeorus* (p. 210)

Fore tarsi longer than fore tibiae; male genitalia not as above (figs. 279, 284, 371) . . . . .Heptageniinae, 42

28(16) Claws of each pair dissimilar (one sharp, one blunt) (fig. 244); costal projection of hind wings acute (fig. 228b); mid-boreal, south to Georgia, Illinois, New Mexico, and California . . . . . .
. . . .*Ameletus* (p. 131)

Claws of each pair similar, sharp (fig. 245); costal projection of hind wings obtuse or weak (figs. 221b, 229b) 29

29(28) Abdominal segments 5–9 of male expanded laterally (fig. 242); tubercles present on sterna of mesothorax and metathorax; rare; mid-boreal, New York . . . .*Siphlonisca* (p. 136)

Abdominal segments 5–9 not expanded laterally; without tubercles on sterna of thorax 30

ADULT CHARACTERS. Fig. 233. Head, *Baetis bicaudatus*. Fig. 234. Hind leg (male), *Pseudiron meridionalis*. Fig. 235. Hind leg (male), *Heptagenia criddlei*. Fig. 236. Hind leg (male), *Siphloplecton basale*. Fig. 237. Apical sterna of abdomen (female), *Pseudiron meridionalis*. Fig. 238. Genitalia (male), *Arthroplea bipunctata*. Fig. 239. Genitalia (male), *Anepeorus simplex*. Fig. 240. Genitalia (male), *Siphlonurus occidentalis*. Fig. 241. Genitalia (male), *Edmundsius agilis*. Fig. 242. Abdomen (male), *Siphlonisca aerodromia*. Fig. 243. Abdomen (male), *Callibaetis nigritus*. Fig. 244. Claws, *Ameletus validus*. Fig. 245. Claws, *Siphlonurus occidentalis*. (Fig. 239 redrawn from Burks, 1953.)

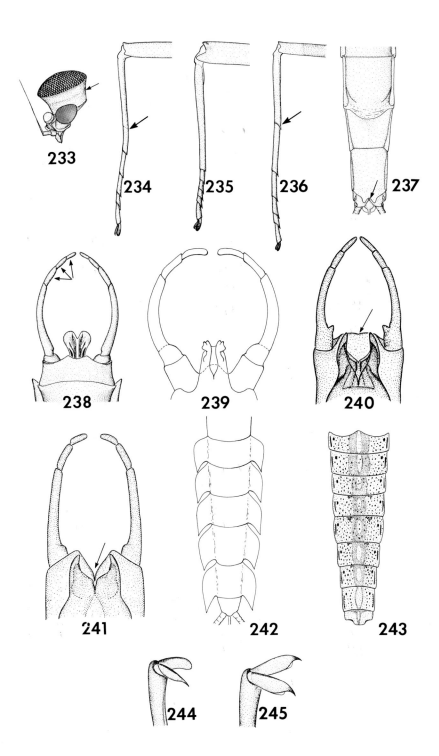

233 234 235 236 237
238 239 240
241 242 243
244 245

30(29)   Vein MP of hind wings simple, unforked (fig. 229b); upper boreal, Canada and western North America south to Wyoming and Utah . . . .*Parameletus* (p. 134)
Vein MP of hind wings forked (fig. 221b) 31

31(30)   Styliger plate of male with V-shaped median emargination (fig. 241); abdominal sterna 3–6 pale with a brown transverse bar; rare; mid-boreal, California . . . .
. . . .*Edmundsius* (p. 133)
Styliger plate of male convex, straight, or slightly concave (fig. 240); ventral markings not as above, usually with spots, longitudinal stripes, oblique stripes, U-shaped marks, or mostly dark; mid-boreal, south to Georgia, Illinois, Arkansas, New Mexico, and California . . . .*Siphlonurus* (p. 138)

32(11,22) Hind wings present although often minute 33
Hind wings absent 37

33(32)   Hind wings with numerous crossveins, usually ten or more; prominent blunt costal projection present on hind wings (figs. 230b, 230c); abdomen with distinct dark speckles similar to those shown in fig. 243; upper austral, widespread north to Alaska . . . .*Callibaetis* (p. 165)
Hind wings with few or no crossveins, usually five or fewer; costal projection of hind wings never blunt, either sharply pointed, hooked, or absent (figs. 231b, c, 232, 247c, 248–252); abdomen without speckled color pattern 34

34(33)   Marginal intercalaries of forewings occur singly (fig. 231a); costal projection of hind wings hooked or recurved (figs. 231b, 231c, 232); mid-boreal, south to northern Mexico . . . .
. . . .*Centroptilum* (p. 169)
Marginal intercalaries of forewings occur in pairs as in fig. 247;

ADULT CHARACTERS. Fig. 246. Forewing, *Cloeon* sp. Fig. 247. (a) Forewing, (b) hind wing, (c) hind wing (enlarged), *Baetis* sp. (*lapponica* group). Fig. 248. (a) Hind wing, *Dactylobaetis cepheus*, (b) same, with shaded arc for comparison with *Baetis*. Fig. 249. (a) Hind wing, *Baetis tricaudatus*, (b) same, with shaded arc for comparison with *Dactylobaetis*. Fig. 250. Hind wing (enlarged), *Heterocloeon*. Fig. 251. Hind wing, *Baetis parvus*. Fig. 252. Hind wing, *Baetis quilleri*. Fig. 253. Forewing, *Baetodes arizonicus*, with shaded arc for comparison with *Pseudocloeon*. Fig. 254. Forewing, *Pseudocloeon* sp., with shaded arc for comparison with *Baetodes*. Fig. 255. (a,b) Marginal intercalary veins, two examples, *Paracloeodes abditus*. Fig. 256. (a,b) Marginal intercalary veins, two examples, *Apobaetis indeprensus*. (Fig. 250 redrawn from Müller-Liebenau, 1974.)

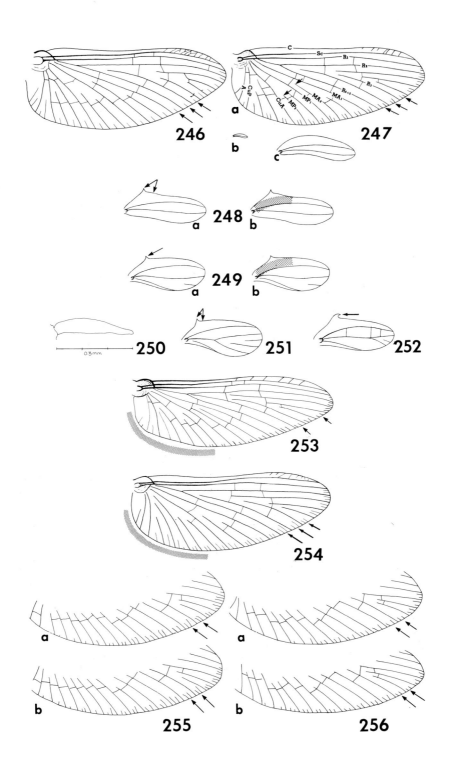

246

247

a

b

c

248 a b

249 a b

0.3mm **250**

251

252

253

254

a

b

255

a

b

256

costal projection of hind wings variable, rarely hooked or absent (figs. 247c, 248–252) 35

35(34)   Costal projection of hind wings broadly based, acute at apex, giving hind wings characteristic shape (fig. 248); mid-austral, Central America north to Oregon, Saskatchewan, and Oklahoma . . . .*Dactylobaetis* (p. 174)

Costal projection of hind wings present or absent (if present, variable in shape); if similar to above, costal projection not broadly based so that hind wings are not shaped as above (figs. 247b, 249–252) 36

36(35)   Hind wings minute in male and female; hind wings without veins (fig. 250); mid-boreal, eastern North America, Quebec, and Ontario south to Georgia . . . .*Heterocloeon* (p. 176)

Hind wings variable, may be minute in female (figs. 247b, 247c); if so, one or two longitudinal veins present; male shown (fig. 398); widespread distribution . . . .*Baetis* (p. 158)

37(32)   Marginal intercalaries of wings occur singly (fig. 246); mid-boreal, south to Idaho, Colorado, Texas, and Florida . . . .
. . . .*Cloeon* (p. 172)

Marginal intercalaries of wings occur in pairs (figs. 253–256) 38

38(37)   Posterior margin of basal half of forewings partly subparallel to costal margin; cubitoanal margin somewhat angular; paired marginal intercalaries short, about as long as width of space between members of a pair (fig. 253); mid-austral, Central America north to Texas, Oklahoma, and Arizona . . . .*Baetodes* (p. 164)

Posterior margin of basal half of forewings rather evenly rounded; cubitoanal margin rounded; paired marginal intercalaries short or long (figs. 254–256) 39

ADULT CHARACTERS. Fig. 257. Mesothorax and metathorax, *Pseudocloeon* sp. Fig. 258. Metathorax, *Paracloeodes abditus*. Fig. 259. Metathorax, *Apobaetis indeprensus*. Fig. 260. Metathorax, *Baetodes* sp. Fig. 261. Mesothorax and metathorax, *Pseudocloeon* sp. Fig. 262. Dorsal edge, mesothorax, and metathorax, *Apobaetis indeprensus*. Fig. 263. Dorsal edge, mesothorax, and metathorax, *Paracloeodes abditus*. Fig. 264. Dorsal edge, mesothorax, and metathorax, *Baetodes* sp. Fig. 265. Head (male), *Epeorus (Ironopsis) grandis*. Fig. 266. Head (male), *Ironodes nitidus*. Fig. 267. Head (male), *Epeorus (Iron) longimanus*. Fig. 268. Penes, *Cinygmula gartrelli*. Fig. 269. Genitalia (male), *Epeorus (Ironopsis) grandis*. Fig. 270. Genitalia (male), *Cinygmula mimus*.

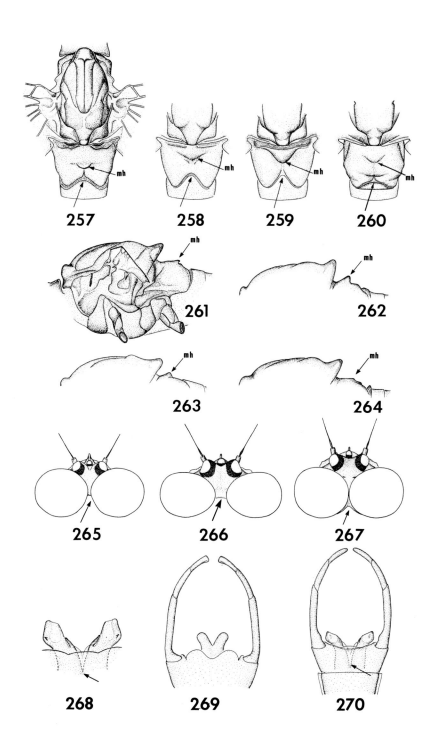

257

258

259

260

261

262

263

264

265

266

267

268

269

270

39(38)   Posterior margin of metanotum shallowly emarginate medially
            (fig. 257); viewed laterally, metascutellar hump (mh) abruptly
            projecting and narrow, flattened dorsally before apex, often
            projecting posteriorly (fig. 261); mid-boreal, south to Florida,
            Tennessee, Illinois, and Utah . . . . *Pseudocloeon* (p. 178)

            Posterior margin of metanotum deeply emarginate medially (figs.
            258, 259); metascutellar hump (mh) broader, not flattened before
            apex or projecting posteriorly (figs. 262, 263) 40

40(39)   Metathorax as in figs. 259, 262; marginal intercalaries short,
            many about as long as width of space between members of a pair
            (figs. 256a, 256b); rare; mid-austral (?), California only . . . . . .
            . . . .*Apobaetis* (p. 157)

            Metathorax as in figs. 258, 263; marginal intercalaries long,
            many distinctly longer than width of space between members of a
            pair (figs. 255a, 255b); mid-austral, north to California, Wy-
            oming, Minnesota, Mississippi, and Georgia . . . .
            . . . .*Paracloeodes* (p. 177)

41(1)   Three caudal filaments present; forewings with only three appar-
            ent longitudinal veins behind vein R_1 and no crossveins (fig.
            196a); mid-austral, Central America north to South Carolina,
            Indiana, Nebraska, and Utah . . . .*Homoeoneuria* (p. 182)

            Two caudal filaments present; forewings with four or more appar-
            ent longitudinal veins behind vein R_1, crossveins present (fig.
            195a); mid-austral, Central America north to Utah, Wyoming,
            and Saskatchewan . . . .*Lachlania* (p. 183)

42(27)   Stigmatic area of wing divided by a fine vein into upper and
            lower series of cellules (figs. 271a, 271c); upper boreal, north-
            western North America south to California, Nevada, Idaho, and
            Wyoming . . . .*Cinygma* (p. 186)

            Stigmatic area of wing with simple crossveins (fig. 273a) or
            anastomosed crossveins (fig. 274a), not divided as above . . .43

43(42)   Stigmatic area of wing with two to many anastomoses of cross-

---

ADULT CHARACTERS. Fig. 271. (a) Forewing, (b) hind wing, (c) detail of stigmatic area,
Cinygma dimicki. Fig. 272. (a) Forewing (with low anastomosis), (b) hind wing, (c) detail of
stigmatic area (with high anastomosis), *Rhithrogena morrisoni*. Fig. 273. (a) Forewing, (b) hind
wing, *Ironodes nitidus*. Fig. 274. (a) Forewing, (b) hind wing, *Epeorus (Ironopsis) grandis*. Fig.
275. (a) Forewing, (b) hind wing, *Epeorus (Iron) albertae*. Fig. 276. (a) Forewing (with high
anastomosis), (b) hind wing, (c) detail of stigmatic area (with low anastomosis), *Cinygmula
mimus*. Fig. 277. (a) Forewing, (b) hind wing, *Stenacron interpunctatum*. Fig. 278. (a) Forewing, (b) hind wing, *Stenonema pudicum*. (Fig. 277 from Jensen, 1974.)

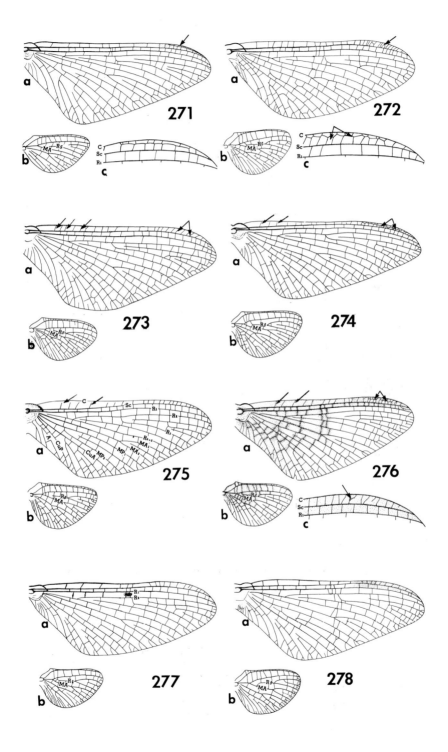

veins (figs. 272a, 272c), and basal segment of fore tarsi one-third or less than one-third length of segment 2; femora usually with dark longitudinal streak near middle; mid-boreal, widespread in North America south to Guatemala . . . . . .*Rhithrogena* (p. 198)

Stigmatic area of wing with or without anastomosed crossveins (figs. 273a, 274a, 276a, c); if crossveins anastomosed (figs. 274a, 276a, c), basal segment of fore tarsi half or more than half length of segment 2; femora usually without dark longitudinal streak 44

44(43)   Basal segment of fore tarsi equal to or slightly longer than segment 2 45

Basal segment of fore tarsi four-fifths or less than four-fifths length of segment 2 47

45(44)   Eyes separated dorsally by more than width of median ocellus (fig. 266); basal costal crossveins of wing well developed, attached anteriorly, stigmatic crossveins strongly slanted (fig. 273a); upper boreal, northwestern North America south to California, Nevada, and Montana . . . .*Ironodes* (p. 197)

Eyes contiguous or nearly contiguous dorsally (figs. 265, 267); basal costal crossveins of wing weakly developed, appearing detached anteriorly, stigmatic crossveins slightly slanted or anastomosed, usually not as above (figs. 274a, 275a); mid-boreal, widespread in North America south to Honduras . *Epeorus*, 46

46(45)   Penes with median titillators minute or absent (fig. 269); stigmatic crossveins of forewings usually anastomosed (fig. 274a); mid-boreal, northwestern North America south to California, Nevada, Wyoming, and Vera Cruz (Mexico) . . . .
. . . . .subgenus *Ironopsis* (p. 193)

Penes with well-developed median titillators (mt) (fig. 279);

ADULT CHARACTERS. Fig. 279. Genitalia (male), *Epeorus (Iron) longimanus*. Fig. 280. Penes, *Stenacron interpunctatum*. Fig. 281. Penes, *Stenacron carolina*. Fig. 282. Penes, *Stenonema sp.* (Utah). Fig. 283. Genitalia (male), *Stenonema nepotellum*. Fig. 284. Genitalia (male), *Heptagenia criddlei*. Fig. 285. Subanal plate (female), *Rhithrogena hageni*. Fig. 286. Subanal plate (female), *Heptagenia solitaria*. Fig. 287. Subanal plate (female), *Cinygmula sp.* Fig. 288. Subanal plate (female), *Epeorus (Ironopsis) grandis*. Fig. 289. Genitalia (male), *Leptophlebia gravastella*. Fig. 290. Genitalia (male), *Paraleptophlebia falcula*. Fig. 291. Genitalia (male), *Ulmeritus carbonelli*. Fig. 292. Penes, *Paraleptophlebia memorialis*. Fig. 293. Penes, *Paraleptophlebia vaciva*. Fig. 294. Penes, *Paraleptophlebia packi*. Fig. 295. Penes, *Paraleptophlebia sp.* (Idaho). (Fig. 279 from Edmunds and Allen, 1964; figs. 280, 281 from Jensen, 1974.)

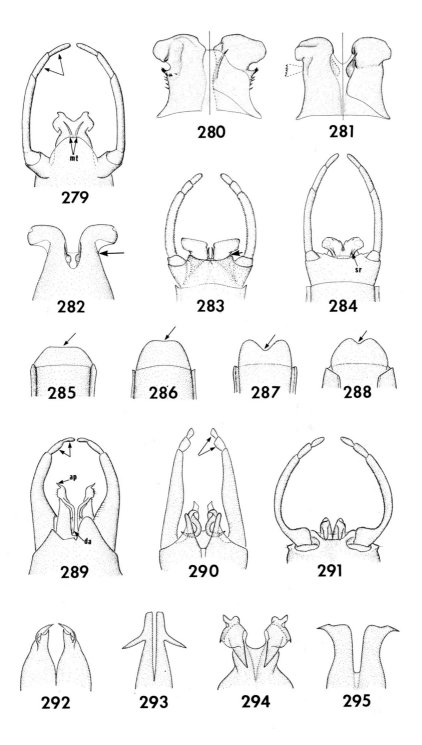

279   280   281   282   283   284   285   286   287   288   289   290   291   292   293   294   295

stigmatic crossveins of forewings not anastomosed (fig. 275a); mid-boreal, widespread in North America south to Costa Rica . . . . .Subgenus *Iron* (p. 192)

47(44)   Penis lobes separated or appearing separated to near base medially (figs. 268, 270); basal segment of fore tarsi usually two-thirds or more than two-thirds length of segment 2; mid-boreal, western North America south to California and New Mexico, eastern North America south to Georgia . . .*Cinygmula* (p. 188)

Penis lobes fused medially at least in basal half (figs. 280–284); basal segment of fore tarsi two-thirds or less than two-thirds length of segment 2, usually less than half length of segment 2 (in *Stenonema lepton*, four-fifths length of segment 2) 48

48(47)   Wings with two or three crossveins below bullae between veins R_1 and R_2 connected or nearly connected by dark pigmentation (fig. 277a), rarely only a dark spot; basal crossveins between R_1 and R_2 dark margined; penes usually with well-developed lateral cluster of spines (fig. 280) (very small in *Stenacron carolina*, fig. 281); mid-boreal, eastern North America south to Florida and west to Arkansas and Minnesota . . . .*Stenacron* (p. 200)

Wings may have crossveins below bullae clouded but never as above (figs. 278a, 296a); crossveins between R_1 and R_2 rarely dark margined; penes without lateral clusters of spines (figs. 282–284) 49

49(48)   Penes distinctly L-shaped, median titillators usually well developed but without subdiscal sclerotized ridge (figs. 282, 283); basal segment of fore tarsi usually one-third to two-thirds length of segment 2; male shown (fig. 404); lower boreal, widespread in North America south to Panama . . . .*Stenonema* (p. 202)

Penes not distinctly L-shaped as above, median titillators moderately to well developed, often with subdiscal sclerotized ridge (sr) present (fig. 284); basal segment of fore tarsi usually one-

---

ADULT CHARACTERS. Fig. 296. (a) Forewing, (b) hind wing, *Heptagenia elegantula*. Fig. 297. (a) Forewing, (b) hind wing, *Metretopus borealis*. Fig. 298. (a) Forewing, (b) hind wing, *Siphloplecton basale*. Fig. 299. (a) Forewing, (b) hind wing, *Leptophlebia pacifica*. Fig. 300. (a) Forewing, (b) hind wing, *Paraleptophlebia debilis*. Fig. 301. (a) Forewing, (b) hind wing, (c) detail of hind wing, *Thraulodes ulmeri*. Fig. 302. (a) Forewing, (b) hind wing, *Ulmeritus carbonelli*. Fig. 303. Hind wing, *Habrophlebia vibrans*. Fig. 304. Hind wing, *Habrophlebiodes americana*.

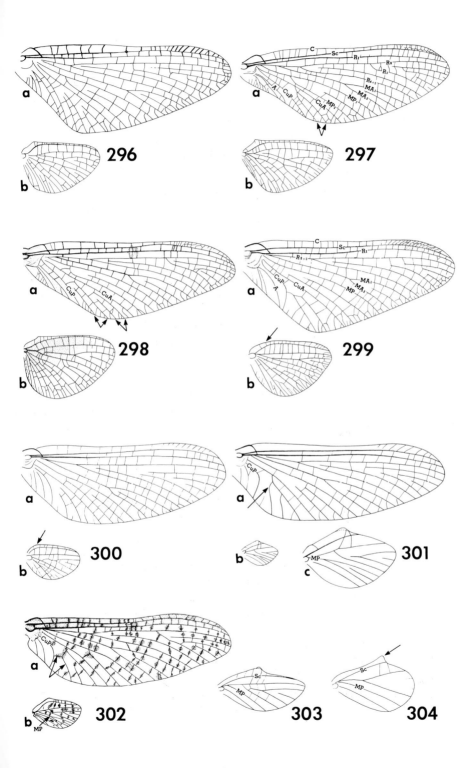

fifth to one-half length of segment 2; lower boreal, widespread south to Guatemala ....*Heptagenia* (p. 194)

50(26)   Stigmatic area of wing divided by fine vein into upper and lower series of cellules (figs. 271a, c); upper boreal, northwestern North America south to California, Nevada, Idaho, and Wyoming ....*Cinygma* (p. 186)

Stigmatic area of wing with simple crossveins (figs. 273a, 277a, 278a, 296a) or anastomosed crossveins (figs. 272a, c, 274a, 276a, c), not divided as above.......................51

51(50)   Subanal plate either broadly rounded or with only slight posteromedian emargination (figs. 285, 286) 52

Subanal plate with moderate to deep V-shaped posteromedian emargination (figs. 287, 288).........................54

52(51)   Stigmatic area of wing with two to many anastomosed crossveins (figs. 272a, c) and femora of most species with dark longitudinal streak near middle; mid-boreal, widespread in North America south to Guatemala ....*Rhithrogena* (p. 198)

Stigmatic area of wing usually without crossveins anastomosed (figs. 277a, 278a, 296a); femora usually without dark longitudinal streak near middle 53

53(52)   Wings with two or three crossveins below bullae between veins R_1 and R_2 connected or nearly connected by dark pigmentation (fig. 277a), rarely only a dark spot; basal crossveins between R_1 and R_2 dark margined; mid-boreal, eastern North America south to Florida and west to Arkansas and Minnesota ....
.................................*Stenacron* (p. 200)

Wings may have crossveins below bullae clouded but never as above (figs. 278a, 296a); basal crossveins between R_1 and R_2 rarely margined; lower boreal, widespread in North America south to Panama. [See discussion of taxonomy under *Heptagenia*.] ....*Heptagenia* (p. 194) and *Stenonema* (p. 202)

ADULTS, MALE GENITALIA. Fig. 305. *Thraulodes* sp. Fig. 306. *Habrophlebia vibrans*. Fig. 307. *Habrophlebiodes brunneipennis*. Fig. 308. *Hermanellopsis incertans*. Fig. 309. *Traverella albertana*. Fig. 310. *Hermanella* sp. Fig. 311. *Ephemerella (Attenella) attenuata*. Fig. 312. *Ephemerella (Drunella) cornutella*. Fig. 313. *Ephemerella (Drunella) doddsi*. Fig. 314. *Ephemerella (Ephemerella) dorothea*. Fig. 315. *Ephemerella (Ephemerella) catawba*. Fig. 316. *Ephemerella (Ephemerella) needhami*. (Fig. 311 from Allen and Edmunds, 1961b; figs. 312, 313 from Allen and Edmunds, 1962b; figs. 314–316 from Allen and Edmunds, 1965.)

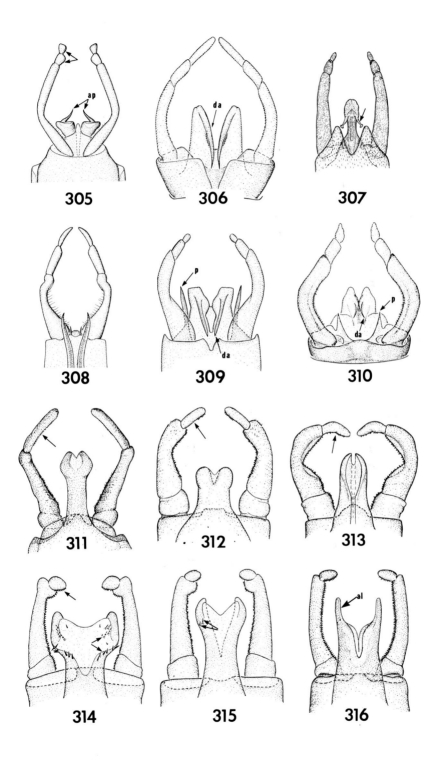

305 **306** **307**

308 **309** **310**

311 **312** **313**

314 **315** **316**

54(51)   Subanal plate with deep V-shaped posteromedian emargination (fig. 287); basal costal crossveins of wing strongly developed (fig. 276a) and posterior margin of head shallowly emarginate; mid-boreal, western North America south to California and New Mexico, eastern North America south to Georgia . . . .*Cinygmula* (p. 188)

                  Subanal plate with shallow to moderate posteromedian emargination (fig. 288); basal costal crossveins of wing usually weakly developed, appearing detached anteriorly (figs. 274a, 275a); if strongly developed (fig. 273a), posterior margin of head deeply emarginate 55

55(54)   Posterior margin of head with deep median emargination; basal costal crossveins of wing strongly developed, attached anteriorly (fig. 273a); upper boreal, northwestern North America south to California, Nevada, and Montana . . . .*Ironodes* (p. 197)

                  Posterior margin of head with only shallow median emargination; basal costal crossveins of wing weakly developed, appearing detached anteriorly (figs. 274a, 275a); mid-boreal, widespread in North America south to Honduras . . . . . .*Epeorus*, 56

56(55)   Stigmatic crossveins of wing usually anastomosed (fig. 274a); wing length 13 mm or greater; mid-boreal, northwestern North America south to California, Nevada, Wyoming, and Vera Cruz (Mexico). .subgenus *Ironopsis* (p. 193)

                  Stigmatic crossveins of wing not anastomosed (fig. 275a); wing length 12 mm or less; mid-boreal, widespread in North America south to Costa Rica . . . . .subgenus *Iron* (p. 192)

57(24)   One pair of cubital intercalaries present (fig. 297a); upper boreal, Canada south to Michigan and Maine . . . .*Metretopus* (p. 148)

ADULT CHARACTERS. Fig. 317. Abdomen, *Ephemerella (Timpanoga) hecuba*. Fig. 318. Genitalia (male), *Ephemerella (Timpanoga) hecuba*. Fig. 319. Genitalia (male), *Ephemerella (Serratella) sordida*. Fig. 320. Abdominal sterna, *Ephemerella (Ephemerella) maculata*. Fig. 321. Genitalia (male), *Ephemerella (Serratella) levis*. Fig. 322. Genitalia (male), *Ephemerella (Ephemerella) maculata*. Fig. 323. Genitalia (male), *Ephemerella (Serratella) tibialis*. Fig. 324. Genitalia (male), *Ephemerella (Dannella) simplex*. Fig. 325. Genitalia (male), *Ephemerella (Eurylophella) lutulenta*. Fig. 326. Genitalia (male), *Campylocia* sp. Fig. 327. Genitalia (male), *Euthyplocia* sp. Fig. 328. Prosternum, *Brachycercus* sp. Fig. 329. Prosternum, *Caenis simulans*. (Fig. 318 from Allen and Edmunds, 1959; figs. 319, 321 from Allen and Edmunds, 1963a; figs. 320, 322 from Allen and Edmunds, 1965; fig. 324 from Allen and Edmunds, 1962a; fig. 325 from Allen and Edmunds, 1963b.)

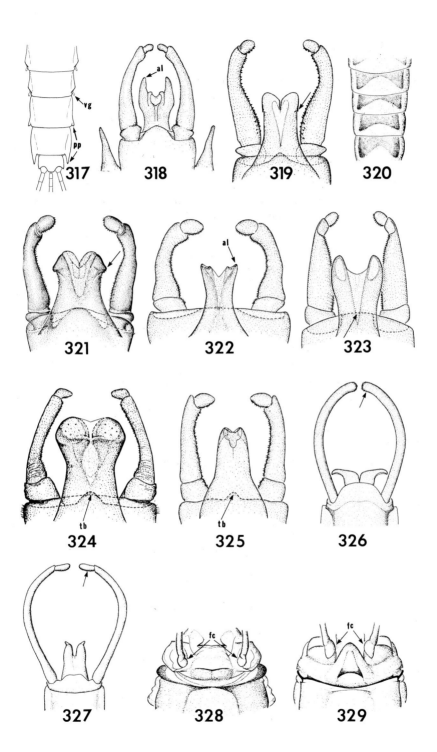

317   318   319   320

321   322   323

324   325   326

327   328   329

Two pairs of cubital intercalaries present (fig. 298a); mid-boreal, Canada south to Illinois and Florida . . . .*Siphloplecton* (p. 149)

58(20)   Hind wings without costal projection (figs. 299b, 300b) 59

Hind wings with distinct costal projection (figs. 301b, 302b, 303, 304, 330b, 331, 332, 333b, 334b) 61

59(58)   Terminal filament shorter and thinner than cerci; penes of male similar to fig. 289, with outer apical projection (ap) and mesal decurrent appendages (da); mid-boreal, south to Oregon, Utah, Illinois, Ohio, and Florida . . . . .*Leptophlebia*, in part, (p. 226)

Terminal filament subequal to cerci; penes of male variable (figs. 289, 290, 292–295) 60

60(59)   Caudal filaments tan with reddish brown ring at articulation of each segment; wing length 8–9 mm; penes of male similar to fig. 289; upper boreal, eastern Canada south to Massachusetts . . . .
. . . .*Leptophlebia*, in part, (p. 226)

Caudal filaments pale or brown, without rings at articulations; wing length variable; penes of male variable (figs. 290, 292–295); mid-boreal, south to Florida, Texas, New Mexico, and California . . . .*Paraleptophlebia* (p. 230)

61(58)   Vein MP of hind wings forked (figs. 301b, c, 302b) 62

Vein MP of hind wings simple, unforked (figs. 303, 304, 330b, c, 331, 332, 333b, c, 334b) 63

62(61)   Second long cubital intercalary vein not attached to CuP basally (fig. 301a); forceps of male close together at base, penes with inwardly directed, spearlike apical projections (ap) (fig. 305); mid-austral, north to Arizona, New Mexico, and Texas . . . . . .
. . . .*Thraulodes* (p. 234)

Second long cubital intercalary vein attached to CuP basally (fig. 302a); forceps of male widely separated, penes without spearlike

ADULT CHARACTERS. Fig. 330. (a) Forewing, (b) hind wing, (c) detail of hind wing, *Choroterpes albiannulata*. Fig. 331. Hind wing, *Choroterpes mexicanus*. Fig. 332. Hind wing, *Hermanellopsis incertans*. Fig. 333. (a) Forewing, (b) hind wing, (c) detail of hind wing, *Traverella albertana*. Fig. 334. (a) Forewing, (b) hind wing, *Homothraulus* sp. Fig. 335. (a) Forewing, (b) hind wing, *Haplohyphes huallaga* (Peru). Fig. 336. (a) Forewing, (b) hind wing (male), *Leptohyphes petersi* Allen (Peru). Fig. 337. Base of forewing, *Haplohyphes mithras*. Fig. 338. (a) Forewing, (b) hind wing, *Euthyplocia*. Fig. 339. (a) Forewing, (b) hind wing, *Ephemera simulans*. (Figs. 335, 337 from Allen, 1966b.)

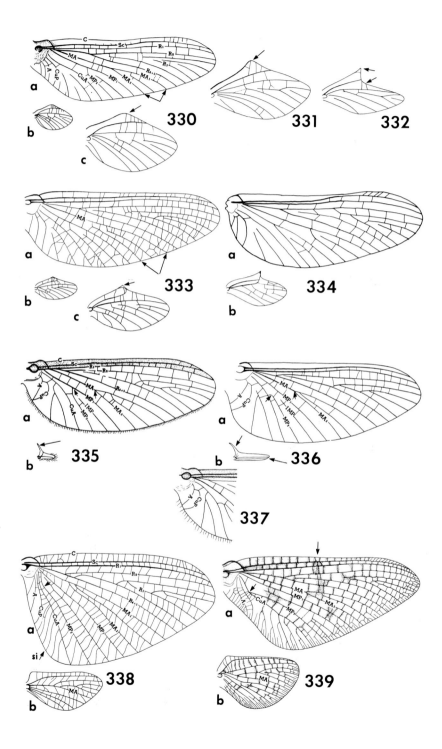

apical projections as above (fig. 291); lower austral, north to Costa Rica .... *Ulmeritus* (p. 238)

63(61)   Vein Sc of hind wings extends well beyond costal projection (fig. 303); each penis lobe of male with a mesal decurrent appendage (da) (fig. 306); mid-boreal, eastern North America south to Florida .... *Habrophlebia* (p. 218)

Vein Sc of hind wings ends at or slightly beyond costal projection (figs. 304, 330b, c, 331, 332, 333b, c, 334b); penes of male variable, not as above (figs. 307–310) 64

64(63)   Costal projection of hind wings in apical half of wing (fig. 304); penes of male with lateral decurrent appendages (da) (fig. 307); mid-boreal, northeastern North America south to Florida, Illinois, and Oklahoma .... *Habrophlebiodes* (p. 218)

Costal projection of hind wing near midpoint of length of wings (figs. 330b, c, 331, 332, 333b, c, 334b); penes of male variable, not as above (figs. 308–310, 375) 65

65(64)   Costal projection of hind wings rounded (figs. 330b, c, 331); fork of vein MA of forewings nearly symmetrical (fig. 330a); lower boreal, widespread south to Panama. [See discussion of taxonomy under *Choroterpes*.] .... *Choroterpes* (p. 214)

Costal projection of hind wings acute (figs. 332, 333b, c, 334b); fork of vein MA of forewings asymmetrical (figs. 333a, 334a) 66

66(65)   Anterior margin of hind wings with extremely large and prominent costal projection (fig. 332); penes of male long and slender (fig. 308); lower austral, north to Darien (Panama) .... *Hermanellopsis* (p. 222)

Anterior margin of hind wings with medium-sized, less prominent costal projection (figs. 333b, 334b); penes of male short and

---

ADULT CHARACTERS. Fig. 340. (a) Forewing, (b) hind wing, *Hexagenia (Hexagenia) limbata.* Fig. 341. (a) Forewing, (b) hind wing, *Hexagenia (Pseudeatonica)* sp. Fig. 342. Head (male), *Litobrancha recurvata.* Fig. 343. Head (male), *Hexagenia (Hexagenia) limbata.* Fig. 344. Mesonotum, *Litobrancha recurvata.* Fig. 345. Mesonotum and metanotum, *Hexagenia (Hexagenia) limbata.* Fig. 346. Genitalia (male), *Litobrancha recurvata.* Fig. 347. Genitalia (male), *Hexagenia (Hexagenia) limbata.* Fig. 348. Genitalia (male), *Hexagenia (Pseudeatonica) callineura.* Fig. 349. Hind leg, *Tortopus* sp. Fig. 350. Hind leg, *Campsurus decoloratus.* Fig. 351. Genitalia (male), left penis revolved to anterior position, *Campsurus decoloratus.* (Figs. 342, 344, 346 from McCafferty, 1971a.)

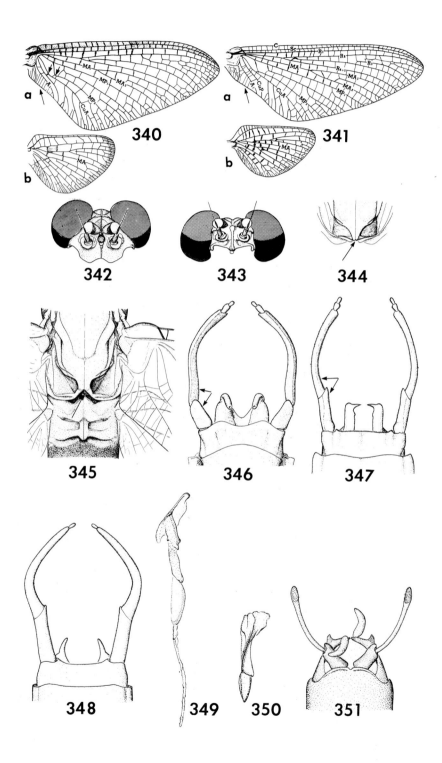

340

341

342 343 344

345 346 347

348 349 350 351

broad (figs. 309, 310). [For species from Texas to Central America, see discussion of taxonomy under *Homothraulus*, p. 224.] 67

67(66)   Each penis lobe of male with long, slender decurrent appendage (da) directed anteromesally; one pair of long to short, stout rod-like projections (p) arises dorsal to forceps (fig. 309); mid-austral, north to Washington, Alberta, Saskatchewan, North Dakota, and Indiana ....*Traverella* (p. 237)

Each penis lobe of male with long, slender decurrent appendage (da) directed anterolaterally; one pair of submedian stout projections (p) arises between forceps (fig. 310); lower austral, north to Honduras ....*Hermanella* (p. 221)

68(20)   Cerci one-fourth to three-fourths as long as terminal filament; upper boreal, northwestern North America south to California, Idaho, and Wyoming ....subgenus *Caudatella* (p. 243)

Cerci and terminal filament subequal in length (unassociated females not keyed beyond this point) 69

69(68)   Terminal segment of genital forceps six times as long as broad (fig. 311); mid-boreal, western North America south to California and New Mexico, eastern North America south to Florida ....subgenus *Attenella* (p. 241)

Terminal segment of genital forceps less than four times as long as broad (figs. 312–316, 318, 319, 321–325)............70

70(69)   Terminal segment of genital forceps more than twice as long as broad; inner margin of long second segment distinctly incurved (fig. 312) or strongly bowed (fig. 313); mid-boreal, western North America south to Baja California and New Mexico, eastern North America south to Michigan and Georgia ....
.........................subgenus *Drunella* (p. 244)

Terminal segment of genital forceps less than twice as long as broad (figs. 314–316, 318, 319, 321–325); inner margin of long second segment not strongly bowed 71

ADULTS, MALE GENITALIA. Fig. 352. Penes, *Ameletus validus*. Fig. 353. *Ameletus sparsatus*. Fig. 354. Penes, *Ameletus oregonensis*. Fig. 355. *Parameletus columbiae*. Fig. 356. *Siphlonisca aerodromia*. Fig. 357. *Analetris eximia*. Fig. 358. *Isonychia sicca campestris*. Fig. 359. *Apobaetis indeprensus*. Fig. 360. *Baetis tricaudatus*. Fig. 361. *Baetodes spiniferum* Traver. Fig. 362. Genitalia and other terminalia, *Callibaetis coloradensis*. Fig. 363. *Centroptilum* sp. (Fig. 357 from Edmunds and Koss, 1972.)

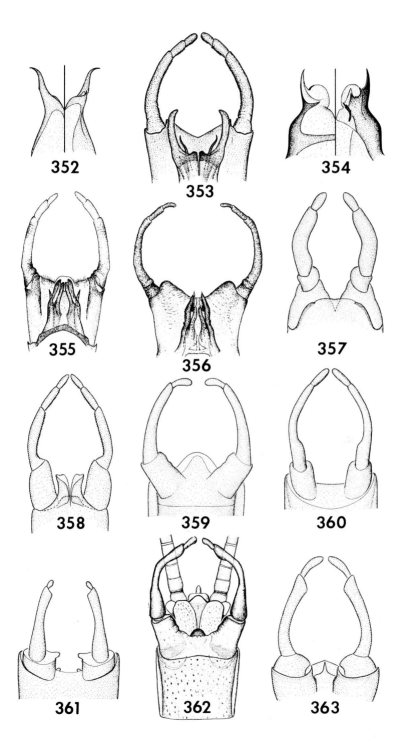

352

353

354

355

356

357

358

359

360

361

362

363

71(70) Penes with dorsal and/or ventral spines, penes usually shaped as in figs. 314, 315; mid-boreal, widespread south to California, New Mexico, Oklahoma, Alabama, and Florida . . . .
. . . . .subgenus *Ephemerella*, in part (p. 247)

Penes without spines, shape variable (figs. 316, 318, 319, 321–325) 72

72(71) Penes with long lateral apical lobes (al) (figs. 316, 318) 73

Penes with shorter lateral apical lobes (al) (figs. 319, 321–325) 74

73(72) Abdomen with well-developed posterolateral projections (pp) on segments 8 and 9 and vestigial nymphal gills (vg) retained on segments 4–7 (fig. 317); penes constricted at base (fig. 318); mid-boreal, western North America south to California and New Mexico .subgenus *Timpanoga* (p. 251)

Abdomen without posterolateral projections and vestigial nymphal gills; penes not constricted at base (fig. 316); mid-boreal, eastern North America south to Missouri and Virginia . . . .
. . . . .subgenus *Ephemerella*, in part (p. 247)

74(72) Abdominal sterna 2–7 with reddish brown to dark brown medially notched rectangular markings (fig. 320); subgenital plate broad, penes small, and genital forceps bowed (fig. 322); mid-boreal, California . . . . .subgenus *Ephemerella*, in part (p. 247)

Abdominal sterna without rectangular markings; subgenital plate narrow, penes variable in size, genital forceps slightly bowed to nearly straight (figs. 319, 321, 323–325) 75

75(74) Penes with lateral subapical projections (figs. 319, 321); mid-boreal, western North America south to Baja California and New Mexico, eastern North America south to Florida, Mississippi, and Arkansas . . . . .subgenus *Serratella*, in part (p. 250)

Penes without lateral subapical projections (figs. 323–325) . .76

ADULTS, MALE GENITALIA. Fig. 364. *Cloeon implicatum*(?) Fig. 365. *Dactylobaetis warreni*. Fig. 366. *Paracloeodes abditus*. Fig. 367. *Pseudocloeon carolina*. Fig. 368. Penes, *Homoeoneuria dolani*. Fig. 369. *Lachlania powelli*. Fig. 370. *Pseudiron meridionalis*. Fig. 371. *Rhithrogena futilis*. Fig. 372. *Metretopus borealis*. Fig. 373. *Siphloplecton basale*. Fig. 374. *Ametropus albrighti*. Fig. 375. *Homothraulus* sp. (Argentina). (Fig. 368 from Edmunds, Berner, and Traver, 1958.)

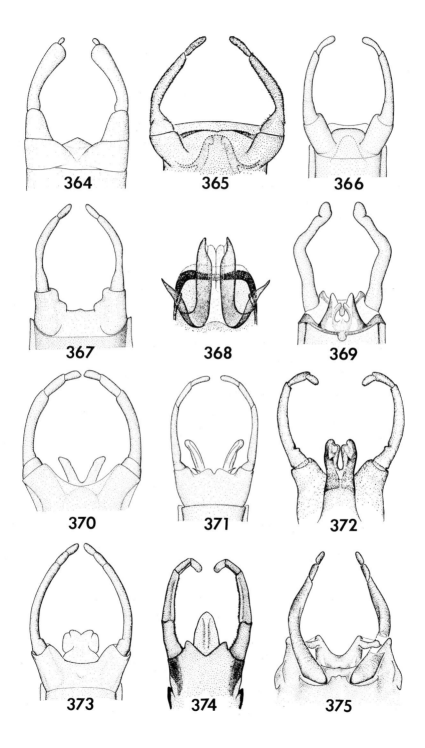

76(75)   Fore tibiae longer than tarsi; subgenital plate without median tubercle (fig. 323); mid-boreal, western United States south to California and New Mexico ....

............................subgenus *Serratella*, in part (p. 250)

Fore tibiae shorter than tarsi; subgenital plate with small median tubercle (tb) (figs. 324, 325) 77

77(76)   Segment 3 of fore tarsi shorter than segment 2; penes relatively narrow apically, expanded basally, long second segment of genital forceps thick (fig. 325); mid-boreal, western North America south to California, eastern North America south to Missouri and Florida ....subgenus *Eurylophella* (p. 248)

Segment 3 of fore tarsi longer than segment 2; penes expanded apically, relatively narrow basally, long second segment of genital forceps thin (fig. 324); mid-boreal, eastern North America south to Illinois and Florida ......subgenus *Dannella* (p. 244)

78(13,18)   Wings of male greatly expanded in cubitoanal areas; vein CuP evenly recurved in male and female (fig. 219); hind wings absent; upper austral, north to British Columbia, Saskatchewan, Quebec, and Newfoundland ....*Tricorythodes* (p. 256)

Wings not expanded in cubitoanal area, broadest near midpoint of length; vein CuP abruptly recurved (figs. 335a, 336a, 337); hind wings present or absent 79

79(78)   Veins CuP and A strongly convergent at wing margin (figs. 335a, 337); hind wings present in male and female; lower austral, north to Costa Rica ....*Haplohyphes* (p. 253)

Veins CuP and A not convergent or only slightly convergent at wing margin (fig. 336a); hind wings present in male, absent in female; mid-austral, north to Utah, Texas, and Maryland .....

...............................*Leptohyphes* (p. 254)

80(13)   Prosternum half as long as broad, rectangular in shape; fore coxae (fc) widely separated on venter (fig. 328); lower boreal,

---

ADULTS, MALE GENITALIA. Fig. 376. *Hagenulopsis* sp. (Brazil). Fig. 377. *Choroterpes mexicanus*. Fig. 378. *Ephemerella (Caudatella) edmundsi*. Fig. 379. *Choroterpes albiannulata*. Fig. 380. *Haplohyphes huallaga* (Peru). Fig. 381. *Leptohyphes mollipes* Traver (Brazil). Fig. 382. *Brachycercus* sp. Fig. 383. *Tricorythodes minutus*. (Fig. 377 from Allen, 1974; fig. 380 from Allen, 1966b.)

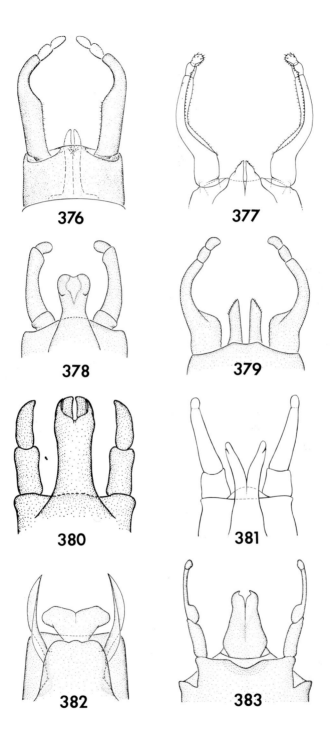

376

377

378

379

380

381

382

383

widespread in eastern and central North America west to Idaho and Utah and south to Panama . . . .*Brachycercus* (p. 264)

Prosternum two to three times as long as broad, triangular in shape; fore coxae (fc) close together on venter (fig. 329); male shown (fig. 422); lower boreal, widespread south to Panama . . . .*Caenis* (p. 266)

81(7)   Wings with one or two intercalary veins behind and parallel to vein CuA; sigmoid intercalaries connect posterior parallel intercalary to hind margin of wing (fig. 201a); genital forceps of male without short terminal segment (fig. 326); lower austral, Central America . . . .*Campylocia* (p. 279)

Wings without parallel intercalaries; sigmoid intercalaries connect vein CuA directly to hind margin of wing (fig. 338a); genital forceps of male with one short terminal segment (fig. 327); lower austral, Central America north to Vera Cruz (Mexico) . . . .*Euthyplocia* (p. 279)

82(9)   Crossveins of wings crowded together near bullae; wings with distinct pattern of dark markings (fig. 339); terminal filament as long as cerci; mid-boreal, south to Utah, Colorado, Oklahoma, and Florida . . . .*Ephemera* (p. 282)

Crossveins of wings not crowded near bullae; wings without pattern of dark markings although crossveins may be darkened (figs. 340a, 341a); terminal filament vestigial or distinctly shorter than cerci 83

83(82)   Head with frons greatly expanded below eyes (fig. 342); scutellum with only shallow medioapical notch (fig. 344); penes of male recurved ventrally (fig. 346); mid-boreal, eastern North America from Ontario and Quebec south to North Carolina . . .
. . . .*Litobrancha* (p. 291)

Head with frons not extending below eyes (fig. 343); scutellum with deep medioapical notch (fig. 345); penes of male bladelike, not recurved (figs. 347, 348); upper austral, Panama north to

---

ADULTS, MALE GENITALIA. Fig. 384. *Neoephemera purpurea*. Fig. 385. *Caenis simulans*. Fig. 386. *Ephemera simulans*. Fig. 387. *Baetisca rogersi*. Fig. 388. *Cinygma integrum*. Fig. 389. *Ironodes nitidus*. (Fig. 387 from Pescador and Peters, 1974.)

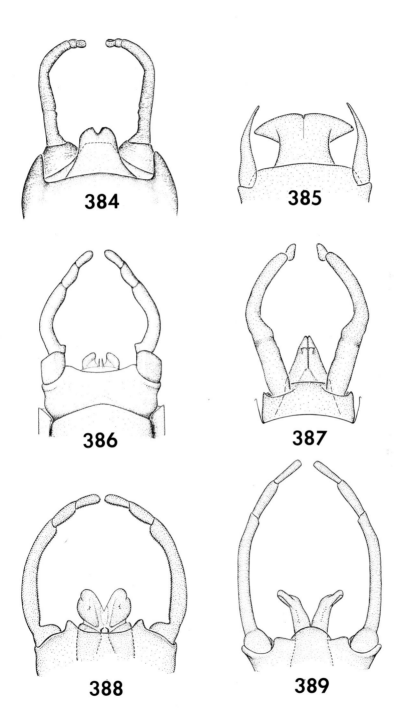

384

385

386

387

388

389

British Columbia, Northwest Territories, Manitoba, Ontario, and New Brunswick ....*Hexagenia*, 84

84(82)   Genital forceps of male with two short terminal segments (fig. 347); eight or more crossveins connect vein A_1 to wing margin (fig. 340a); male shown (fig. 428); upper austral, Jalisco (Mexico) north to British Columbia, Northwest Territories, Manitoba, Ontario, and New Brunswick ......subgenus *Hexagenia* (p. 288)

Genital forceps of male with one short terminal segment (fig. 348); six or fewer crossveins connect vein A_1 to wing margin (fig. 341a); lower austral, Central America north to southern Mexico ....subgenus *Pseudeatonica* (p. 290)

85(6)   Middle and hind legs with all segments present, but reduced and shriveled (fig. 349); clawlike appendage extending laterally from bases of forceps of male (figs. 206, 207); mid-austral, Central America north to Nebraska, Manitoba, South Carolina, and Florida ....*Tortopus* (p. 300)

Middle and hind legs greatly reduced, terminating in a bladelike femur (fig. 350); without clawlike appendage at bases of forceps (fig. 351); mid-austral, Central America north to Texas ......
...............................*Campsurus* (p. 298)

The Families of Mayflies

In the familial accounts distinctive taxonomic characters are given for each family, the superfamily assignment is noted, and the most probable cladistic relationships are indicated. For each family and subfamily the total world distribution is discussed. Familial and subfamilial diagnostic characters are given for nymphs and adults of each family and subfamily. Brief accounts are given for the three families not found in the Americas. Within each subfamily the discussions of North and Central American genera and subgenera cover nymphal characteristics, nymphal habits and habitats, life history, adult characteristics, mating flights, taxonomy, and distribution. Reference is made to all nymphal and adult features illustrated in the text.

Nymphal Characteristics. The diagnostic characters listed will serve to distinguish the nymphs of each genus or subgenus from those of other genera in the family or other subgenera in the genus.

Nymphal Habitats. The descriptions under the generic accounts are largely anecdotal. The value of the accounts is somewhat limited because of the diversity of habitats for a single genus. The habitats of the smaller genera may be easily generalized, but the larger the genus, the more difficult it is to generalize the habitat. For example, in the Rocky Mountains most of the species of *Rhithrogena* live in medium to large trout streams, but *R. robusta* departs from this pattern and lives in extremely cold headwater streams and *R. undulata* is found only in silted warm streams. Within a stream the nymphs of various ages in a single species often gradually shift to different microhabitats as they mature. The nymphs of some leptophlebiids migrate into small tributaries or backwaters when they start to mature, and lake-inhabiting *Leptophlebia* and *Paraleptophlebia* move from deep water toward the shallow areas near shore or become isolated in small residual pools just before emergence. All of these factors combine to make generalizations on the nymphal habitats of the genera very broad, especially for genera with many species.

123

There is a considerable need to synthesize the information on aquatic insect habitats. The partitioning of any single aquatic stream, pond, or lake by insects is poorly understood but probably is largely done in three ways — spatial arrangement, seasonal partitioning, and specialization for food resources available at a given time. Any one species can of course inhabit only a small portion of the aquatic environment. Each species is limited by such physical and chemical parameters as temperature regime, oxygen, current, turbidity, and extremes of other factors.

The mayflies are generally microhabitat specialists. In ponds and lakes they are clearly limited by substrate, depth, and wave action. In streams substrate and current combine to create many subdivisions in which two microhabitats may be separated by only a few centimeters.

Most stream sections have only a small part of their total species diversity present at one time. For example, in the Green River of Utah Edmunds and Musser (1960) and Edmunds (unpublished) found the spring biomass to be dominated by *Leptophlebia gravastella, Ephemerella inermis,* and *Isonychia sicca*; the autumn biomass was predominantly *Lachlania powelli, Traverella albertana*, and *Heptagenia elegantula.* Frequently two or more species of a single genus seasonally succeed each other in a single stream; for example, *Epeorus longimanus* is followed by *E. deceptivus* and *Cinygmula mimus* is followed by *C. par* and then *C. tarda* in Utah streams. Seasonal spatial separation of related species of similar size need not be complete to be effective, because different size classes of the same or related species often have different microhabitat and food requirements.

Nymphal Habits. There have been few detailed studies of the nymphal habits of mayflies. Furthermore, most of the data are anecdotal and the result of casual observations. Even so, such data can prove valuable in many ways. The nymphs of all Siphlonuridae, for example, climb out of the water before the subimagoes emerge; elsewhere in the order the phenomenon is rather rare but is found in primitive members of several lineages. Such data are often of use in determining phylogenetic relationships. Although nymphal habits tend to be fairly uniform within a genus, there are exceptions, especially in the large genera.

Mayflies feed largely as opportunistic generalists. Although a species may be primarily herbivorous, it will feed readily on detritus or animal material. Especially when stressed or confined to abnormal areas mayflies readily adapt their feeding habits to whatever material is available.

Cummins (1973) divides aquatic insects into four general trophic categories based on feeding mechanisms: shredders, collectors, scrapers, and predators. Each of these categories has two subdivisions, thus making eight categories, of which five apply to mayflies. None of the shredders Cummins lists are mayflies, and we know of no mayflies that should be placed in that category. Among the filter-feeding or suspension-feeding collectors the only mayflies Cummins lists are the Siphlonuridae. Presumably Cummins refers only to *Isonychia* (Isonychiinae). To this category should be added *Lachlania* and some Leptophlebiidae (*Traverella* and *Hermanella*). The burrowers found in hard clay river walls that are swept by the currents (*Ephoron*, *Campsurus*, and *Tortopus*) should be placed here, and much of the feeding by *Hexagenia* is also filter-feeding. Most mayflies are collectors on sediments or feed on fine detritus in surface deposits. This feeding habit applies probably to all Siphlonurinae (except that most *Ameletus* are probably largely scrapers), most Baetidae, various Heptageniinae (especially some *Stenonema*, *Stenacron*, and *Heptagenia*), a large part of the Leptophlebiidae (except *Hermanella* and *Traverella*), many *Ephemerella*, the Tricorythidae in large part, some *Ephemera*, and the Caenidae and Baetiscidae. Mayflies that live in U-shaped burrows (for example, *Hexagenia* and *Ephoron*) on the bottom frequently leave the burrows to feed on surface deposits. The remarkably adapted sand-burrowing genus *Homoeoneuria* collects algae and detritus from shifting sand, and *Ametropus* feeds in the same way on the upper centimeter of stable sand surfaces.

The mineral scraper collectors of periphyton include *Ameletus*, a number of Baetidae, most Heptageniinae, and a number of Ephemerellidae and Leptophlebiidae. The organic scrapers of algae and associated periphyton would also include some *Ameletus*, many Baetidae, some Heptageniidae, many Leptophlebiidae, some Tricorythidae, and most Caenidae.

To some degree many mayflies take in a small number of animals while feeding; for instance, *Isonychia*, *Siphlonurus*, and *Ephemera* may eat considerable quantities of animal material. But several genera are true predators. The sand-dwelling siphlonurid genus *Analetris* and the rock-inhabiting heptageniid *Anepeorus* are both known to feed on chironomid larvae; this is probably true also of *Acanthametropus*. The unusual heptageniid genera *Pseudiron* and *Spinadis* also feed on chironomids, and the sand-burrowing behningiid *Dolania* consumes larvae of chironomids and ceratopogonids.

Most mayflies feed on particles of various sizes, according to the size and adaptive mode of the nymphs. Mature nymphs of *Siphlonurus occidentalis* prey extensively on tube-dwelling chironomid larvae, but they will feed just as readily on mosquito larvae, plant fragments, or fine gravel of the same size. When the mayfly nymphs of this species are small, they grow well on much finer material. The habit of eating increasingly larger particles as the nymph increases in size is a common pattern in Ephemeroptera.

The general feeding patterns can be summarized for the various taxonomic groups only if it is realized that each species generally feeds on relatively small particles when the nymphs are young and that the large species generally feed on larger particles than do small species of the same genus. The accompanying graph shows the relative percentage of detritus, algae, and animal matter fed upon by *Stenonema fuscum* nymphs of various sizes (from Coffman et al., 1971).

Life History. Relatively complete life history data are rare for North and Central American Ephemeroptera. Almost certainly there is a tendency to overestimate the length of time needed for nymphal development. There has also been a tendency to infer, for example, that the existence of three size classes of nymphs means that nymphal development takes three years. Such assumptions are dangerous and may lead to errors. The eggs from a single female may hatch over a long period of time, or differences in size may exist because of differential availability of food. The time necessary for development from egg to adult varies considerably with temperature within a single species. *Hexagenia limbata*, for example, may take two years to develop in a northern lake (Neave, 1932) and only about seventeen weeks in canals in

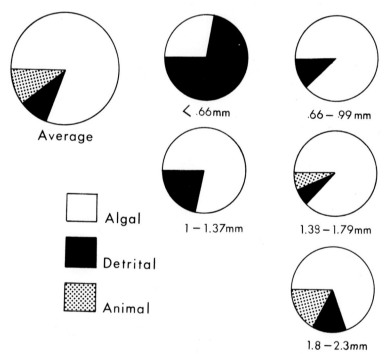

Food of *Stenonema fuscum* nymphs, showing average of all sizes, and change in diet with increased size as indicated by range of head capsule width in mm. (Redrawn from Coffman, Cummins, and Wuycheck, 1971.)

Utah. Within a single species there may be one, two, or more broods per year, depending on the temperature.

Landa (1968) has suggested the following life-history classification for Ephemeroptera, based on the Central European species. There are four groups and eleven types.

GROUP A. One generation per year.

> Type A1. Eggs hatch in autumn; nymphs continue to grow during winter; adults emerge in spring or summer.
>
> Type A2. Eggs in diapause until spring or summer; growth of nymphs is short and rapid.
>
> Type A3. Nymphs hibernate over winter without growth and resume development in spring and summer.

GROUP B. Two generations per year.

> Type B1. First-generation eggs hatch in autumn; nymphs develop during winter; adults emerge in spring; second generation develops during summer.
>
> Type B2. Eggs in diapause until spring or summer; two generations develop in succession in summer.
>
> Type B3. First-generation eggs hatch in autumn; adults emerge in spring; two generations develop in succession in summer.
>
> Type B4. Some eggs hatch in autumn with nymphs becoming well developed, other eggs hatch in spring; adults emerge as two broods in summer.

GROUP C. Species with two-year or three-year development period.

> Type C1. Development in two years.
>
> Type C2. Development in three years.

GROUP D. Species with three generations in two years or two generations in three years.

> Type D1. Three generations in two years.
>
> Type D2. Two generations in three years.

This life-history classification can be applied to the North American and Central American mayflies with some modifications. Obviously the Central American seasonal pattern of wet and dry seasons with small temperature shifts cannot be accommodated by this system. Nevertheless, a large part of the tropical and subtropical mayfly fauna appears to be seasonal.

As Landa notes, a species may shift from one group or division to another according to temperature changes with altitude and latitude. Type B4 is not reported from North America. The only North American species with a two-year life cycle (type C1) is *Hexagenia limbata* in the northern part of its range. Type C2 (three years) is reported but not confirmed in Europe and is not reported in North America or Central America. Type D2 (two generations in three years) also is not confirmed in Europe and is not reported in the Americas.

Most of the species treated in this volume are of life-history types A1 or A2, with a number of species, principally Baetidae and Caenidae, of type B1. Type B2 is represented in North America by *Siphlonurus occidentalis, Caenis simulans*, and a number of other species. *Callibaetis nigritus* and probably several others are representatives of type B3. Type D1 was described by Murphy (1922) for *Baetis vagans* McDunnough, but it probably applies to a number of other *Baetis* species in suitable temperature regimes. Hynes (1970) has a simpler annual cycle classification based on nonseasonal, slow seasonal, and fast seasonal patterns.

It is probable that a number of mayflies, especially some of the Baetidae and Caenidae species, become nonseasonal in southern climates. Some species that are normally seasonal in most streams may become aseasonal in spring-fed streams or below dams where similar "flat" temperature regimes (in which there is little difference between summer high and winter low temperatures) occur. Even in the strongly seasonal climate of Salt Lake City, Utah, *Baetis tricaudatus* subimagoes emerge during warm afternoons in mid-winter, but these emergences are minor compared to the spring emergence.

Slow seasonal cycles apply to many mayflies, including several species of *Ephemerella*. Many other examples are cited in the life-history accounts. The slow seasonal type encompasses A1, A3, C1, C2, D1, and D2 of Landa's system.

Fast seasonal types are exemplified by most Baetidae, Caenidae, and various representatives of other families. The group includes types A2, B1, B2, B3, and B4 of Landa. An extreme case of fast development in mayflies is *Parameletus columbiae*, whose nymphal development in Utah is complete in sixteen to twenty-two days.

Our knowledge of the seasonal distribution of most species is incomplete, and general statements for the genera may be of little value. Many of the records coincide with the period of maximum field activity of biologists. In areas presumed to be well worked, additional species may be found early and late in the season. Even in the northern states some species emerge only in late fall or early winter; in northern Utah, for example, two species of *Ameletus* emerge mainly from October through December. Also in Utah, *Parameletus columbiae* is common in certain high-altitude localities only for sixteen to twenty-two days after the ice melts and spends the rest of the year in the egg stage.

Because of the role of temperature seasonal emergence is highly variable. In general, species emerge earlier in the south than in the north, but occasionally the reverse is true. Areas with mild winter temperatures, such as the coastal areas of California, Oregon, Washington, and British Columbia and the southern United States, Mexico, and Central America, tend to be areas of early emergence.

Species with long emergence periods may be taxonomically difficult because of the influence of temperature on size and color. Most species are smaller and paler when they develop at higher temperatures.

Adult Characteristics. Diagnostic characters are given to distinguish the adults of each family, subfamily, genus, and subgenus from other members of the taxon in which they are included.

Mating Flights. Data for each genus on the time and the location of mating swarms may have limited value. Because the time and the place of swarming often serve to isolate related species, it is to be expected that generalizations about large genera may also be of little value. In *Paraleptophlebia*, *Siphlonurus*, and probably other genera two species may swarm at quite different times of day. In *Callibaetis*, however, a consistent pattern may be seen for several species in a single area. Such species as *Paraleptophlebia debilis* fly at higher light intensities as the temperature drops. The swarming of some genera is enigmatic because either the time or the place has escaped the attention of biologists. Some species swarm at night and others swarm at dawn when few biologists are afield. A considerable number of species swarm at distances as much as one mile from their nymphal home, and a number of species of *Ephemerella*, and probably other genera as well, swarm fifty to one hundred feet (or more) above the ground.

Taxonomy. For each genus we give a reference to the original description and cite the type species and its geographic origin. The discussion attempts to evaluate the present state of taxonomy for species identification. In a few cases certain species problems are discussed in detail. If the best keys are in other than standard works such as Needham, Traver, and Hsu (1935), Berner (1950), or Burks (1953), we give references to the current keys.

Distribution. The total distribution and the North American and Central American distribution are given for each genus and subgenus. The distribution of the species of each genus or subgenus over the twelve geographic areas of North and Central America (indicated on the map on page 50) is noted in the discussion or is shown on a distribution chart. The author's name following each species is a useful lead to the literature on the species of the genus. Parentheses around the author's name indicate that the species was described in a genus other than the one where it now is found.

Family SIPHLONURIDAE

The Siphlonuridae of the superfamily Heptagenioidea exhibit more primitive characters than any other extant family of mayflies. The family appears to be an assemblage of forms from an early adaptive radiation in the order. Most of the families of Heptagenioidea can be traced from the Siphlonuridae, and some of the characters of the south temperate subfamily Oniscigastrinae suggest that it is near the base of the phyletic line that leads to the non-heptagenioid families.

Most of the genera of the family are adapted to cool waters and are most diverse in the Holarctic region and in Australia, New Zealand, and southern South America. Four subfamilies — Oniscigastrinae, Coloburiscinae, Ameletopsinae, and Rallidentinae — are confined to the south temperate regions. Three subfamilies — Siphlonurinae, Acanthametropodinae, and Isonychiinae — are present in the Nearctic with some Siphlonurinae also in the south temperate regions. The descriptions have been formulated primarily for the North and Central American genera.

Nymphal Characteristics. Body form generally streamlined and minnow-like. Head hypognathous, eyes lateral, antennae relatively short (rarely exceeding twice width of head). Gills on segments 1–7. Labium with glossae and paraglossae short and broad. Abdomen somewhat flattened, the posterolateral angles of the abdominal segments, especially the apical ones, expanded into flat lateral extensions terminating in posterolateral projections (poorly developed in some species of *Ameletus*). Three caudal filaments; terminal filament with dense row of setae on each side; cerci with dense setae on mesal side only.

Adult Characteristics. Eyes of male relatively simple, upper portion usually with larger facets than the lower portion. Wings narrow and triangular; hind wings quite large. Veins and crossveins numerous in both pairs of wings. Vein CuA attached to hind margin of wing by a series of veinlets, often forking or sinuate. Forks of veins MP and CuA relatively symmetrical, MP_2 and CuA not divergent from MP_1 basally.

Subfamily Siphlonurinae

Members of this subfamily are found in the temperate northern and southern hemispheres, being most diverse in the Holarctic region.

Nymphal Characteristics. Body always minnowlike. Antennae shorter than twice width of head. Mouthparts adapted for general feeding; mandibles with distinct molar area; maxillae and labium with three-segmented palpi.

Adult Characteristics. Eyes of male with upper portion of larger facets. Hind wings less than half as long as forewings. Vein MP of hind wings simple and unforked, or forked near base. Two caudal filaments; terminal filament a short stub.

Ameletus Eaton (figs. 63, 76–78, 228, 244, 352–354, 390)

Nymphal Characteristics. Body length: 6–14 mm. Crown of galea-lacinia of maxillae with many long pectinate spines. Posterolateral projections on apical abdominal segments poorly to moderately developed. Gills single, oval, and small, with lateral and mesal or submesal sclerotized bands. Caudal filaments usually with a wide dark transverse band at middle and a narrow dark band at apex or with many small dark bands. Illustration: Fig. 390.

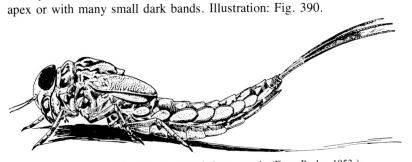

Fig. 390. *Ameletus ludens*, nymph. (From Burks, 1953.)

Nymphal Habitat. The nymphs are usually found in small, rapidly flowing streams among pebbles, near banks, among vegetation and debris, or often in rocky pools on the sides of boulders. In the Rocky Mountains some species are abundant in tiny streams that are only a few inches wide and less than one inch deep. The nymphs are sometimes found in large rivers, lakes, and ponds with clean rock or gravel bottoms. Occasionally they are found in streams at altitudes as high as 11,000 feet.

Nymphal Habits. Strong swimmers; the nymphs can swim in water flowing at the rate of two to three feet per second, but they seek quieter water before coming to rest on the bottom. One species has been observed only between and behind small stones at the water's edge, where the nymphs were well protected from the slightest current.

Life History. Development from egg to adult probably takes one year or less. The European specis *A. inopinatus* oviposits in the summer, and the eggs hatch after a period of about one month, reaching the tenth-instar nymphal stage in autumn. The nymphs overwinter and emerge in the spring or summer months; the number of instars is unknown. Emergence occurs as early as February in the southeast and as late as September in Canada. In Utah *A*.

validus emergence dates range from late October to late December. *Ameletus similior* also emerges in late autumn. The nymph leaves the water, usually going to the top of a rock or several inches up a twig, and remains quiescent for up to fifteen minutes before the subimago emerges. If disturbed, the nymph flips over backward into the stream. The eggs are laid a few at a time in scattered locations. The eggs of some species are laid in streams which become dry during part of year, hatching after the stream flow resumes.

Adult Characteristics. Wing length: 8–14 mm. Usually brownish or yellowish brown; coloration of male and female similar with abdomen of female often darker than that of male. Eyes of male large, contiguous dorsally. Forelegs of male nearly as long as body; basal tarsal segment one-third to one-half length of segment 2. Claws dissimilar. Hind wings with costal projection acute. Male genitalia variable, examples illustrated in figs. 352–354.

Mating Flights. Not described. During three days in the field on a stream in Utah where large numbers of male and female subimagoes had emerged, no swarming was detected at any time. The males may swarm high in the air or at a site remote from the water, or they may swarm at night. *Ameletus ludens* has been reported to be parthenogenetic.

Taxonomy. Eaton, 1885. *Trans. Linn. Soc. London*, 2d Ser. Zool., 3:210. Type species: *A. subnotatus* Eaton; type locality: Colorado. Specific identification of the nymphs of a few species may be made if caution is used. Even the males are often difficult to identify to species, principally because adults have seldom been collected in large numbers to determine the variation. Females cannot be identified unless associated with males.

Distribution. Holarctic. In Nearctic, most abundant and diverse in the north, extending along the mountains south to California, New Mexico, Illinois, and Georgia.

Species	United States						Canada			Mexico		Central America
	SE	NE	C	SW	NW	A	E	C	W	N	S	
aequivocus McDunnough	...	...	X	...	...	...	...	...	...	...	...	...
alticolus McDunnough	...	...	...	...	...	...	...	...	X	...	...	...
amador Mayo	...	...	...	X	...	...	...	...	...	...	...	...
browni McDunnough	...	...	...	...	...	X	...	...	...	...	...	...
celer McDunnough	...	...	...	...	...	...	...	...	X	...	...	...
celeroides McDunnough	...	...	...	...	...	...	...	...	X	...	...	...
connectina McDunnough	...	...	...	...	...	...	...	...	X	...	...	...

Species	United States						Canada			Mexico		Central America
	SE	NE	C	SW	NW	A	E	C	W	N	S	
connectus McDunnough	...	...	...	...	X	...	...	...	X	...	...	...
cooki McDunnough	...	...	...	...	X	...	...	...	...	...	...	...
dissitus Eaton	...	...	...	X	...	...	...	...	...	...	...	...
exquisitus Eaton	...	...	...	...	X	...	...	...	...	...	...	...
facilis Day	...	...	...	X	...	...	...	...	...	...	...	...
falsus McDunnough	...	...	...	X	...	...	...	...	...	...	...	...
imbellis Day	...	...	...	X	...	...	...	...	...	...	...	...
lineatus Traver	X	...	X	...	...	...	...	...	...	...	...	...
ludens Needham	...	X	X	...	...	...	...	...	...	...	...	...
monta Mayo	...	...	...	...	...	...	...	...	X	...	...	...
oregonensis Eaton	...	...	...	...	X	...	...	...	X	...	...	...
querulus McDunnough	...	...	...	...	...	...	...	...	X	...	...	...
shepherdi Traver	...	...	...	X	...	...	...	...	...	...	...	...
similior McDunnough	...	...	...	X	X	...	...	...	X	...	...	...
sparsatus McDunnough	...	...	...	...	X	...	...	...	X	...	...	...
subnotatus Eaton	...	...	X	X	...	...	...	...	X	...	...	...
suffusus McDunnough	...	...	...	...	X	...	...	...	X	...	...	...
tertius McDunnough	...	...	...	...	...	...	X	...	...	...	...	...
tuberculatus McDunnough	...	...	...	...	...	...	...	...	X	...	...	...
validus McDunnough	...	...	...	X	X	X	...	...	X	...	...	...
vancouverensis McDunnough	...	...	...	...	...	...	...	...	X	...	...	...
velox Dodds	...	...	...	X	X	...	...	...	X	...	...	...
vernalis McDunnough	...	...	...	...	...	...	...	...	X	...	...	...
walleyi Harper	...	...	...	...	...	...	X	...	...	...	...	...

Edmundsius Day (figs. 73, 74, 241)

Nymphal Characteristics. Body length: 15–17 mm. Frontal margin of labrum broadly and deeply excavated. Claws of middle and hind legs almost twice the length of claws of forelegs. Posterolateral projections on the apical abdominal segments moderately well developed. Gills double on abdominal segments 1–2, with ventral lamellae about two-thirds as large as dorsal lamellae, both lamellae oval; other gills regularly oval, single. Illustration: Not illustrated in full; parts in Day, 1956, fig. 3:27, m, q.

Nymphal Habitat. Slow-flowing, shallow, well-aerated streams. The nymphs rest close to shady edges of pools on sand or fine gravel bottoms. They are found in streams at elevations of 5,000 to 8,000 feet.

Nymphal Habits. Not described.

Life History. Emergence occurs between sunset and sunrise during July and early August. Duration of nymphal life unknown. The nymphs crawl out of the water for the subimagoes to emerge.

Adult Characteristics. Wing length: 16 mm. Eyes of male contiguous dorsally, with lower portion darker. Forelegs of male about two-thirds length of body; tarsi nearly 2⅓ times length of tibiae. Claws similar. Hind wings with costal projection rounded. Male genitalia illustrated in fig. 241.

Mating Flights. Unknown.

Taxonomy. Day, 1953. *Pan-Pac. Entomol*. 29:19. Type species: *E. agilis* Day; type locality: Willow Creek, Madera County, California. The single species is known only in California.

Distribution. Known only from Madera County, California.

Parameletus Bengtsson (figs. 229, 355, 391)

Nymphal Characteristics. Body length: 10–13 mm. Apex of the labial palpi pincerlike. Claws long and slender. Posterolateral projections on abdominal segments well developed. Gill lamellae broad, cordate, single on all segments. Illustration: Fig. 391.

Nymphal Habitat. Primarily in swamps and forest pools with emergent vegetation, especially *Carex*. The nymphs are most abundant where there is a barely perceptible current. The water temperature of pools may exceed the air temperature in the middle of the day.

Nymphal Habits. The nymphs feed along the bottom or on stems of *Carex*. The gills are vibrated intermittently. The nymphs move slowly unless disturbed, then dart five to ten inches away and settle to the bottom again. Disruptive color pattern.

Life History. In Utah the eggs are deposited in mid-June and remain dormant until they hatch in the following May or June, within one day after the snow melts. Six days after hatching the nymphs are 4 or 5 mm in length, and after eight days they reach 5 or 6 mm. Nymphal life is completed in sixteen to twenty-two days. The nymph crawls one or two inches out of the water for the subimago to emerge. Emergence occurs throughout the day, but more subimagoes emerge in the morning hours than later in the day.

Adult Characteristics. Wing length: 10–13 mm. Brownish; male and female very similar in coloration. Eyes of male large, contiguous dorsally. Forelegs of male as long as body; basal segment of fore tarsi and fore tibiae thickly set with blunt, spinelike processes; spines at tips of tarsal segments of middle and hind legs; claws similar. Forewings with stigmatic crossveins anastomosed. Hind wings usually with a low, broadly rounded, costal projection. Male genitalia illustrated in fig. 355.

Mating Flights. Brinck (1957) has reported observations in Europe of the

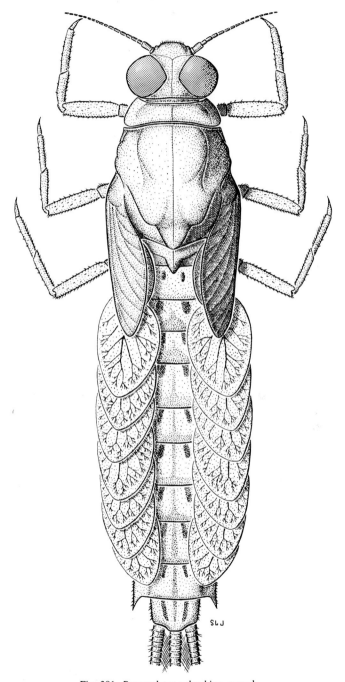

Fig. 391. *Parameletus columbiae*, nymph.

Holarctic species, *P. chelifer* Bengtsson. This species swarms in the evening or early in the night. Initially the swarm contains mostly males, but soon numerous females join the flight. The male approaches the female from beneath, extends the forelegs upward and forward, and curves the fore tarsi posteriorly until they are caught at the wing bases of the female. At the same time the forceps of the male grasp the posterior end of the abdomen of the female, which is recurved, and the penes are inserted at the posterior margin of segment 7. Copulation is completed in about twenty seconds. While *in copula* the insects lose height but always separate before reaching the ground or the water surface. Many couples mate in the air, but Brinck (1957) noted that a considerable part of the population alighted on a bridge where the males crept to the females and succeeded in copulating.

Taxonomy. Bengtsson, 1908. *Vet. Akad. Arsbok.* 6:242. Type species: *P. chelifer* Bengtsson; type locality: Bjorkfors, Sweden. The males of *Parameletus* should be readily distinguishable to species by the characters of the genitalia. No key to the nymphs is available.

The generic name *Parameletus* Bengtsson appears to be a *nomen nudum*, or name without description. If so, the valid name of this genus is *Sparrea* Esben-Peterson. The name *P. chelifer* Bengtsson would also be replaced by *S. norvegica* Esben-Peterson. We are not changing the name until we have sought an opinion on the conservation of the names *Parameletus* and *P. chelifer* Bengtsson.

Distribution. Holarctic, generally confined to northern regions. In the Nearctic, in Alaska and Canada and south through the Rocky Mountains to high elevations in northern Utah.

Species	United States						Canada			Mexico		Central America
	SE	NE	C	SW	NW	A	E	C	W	N	S	
chelifer Bengtsson	...	...	...	...	...	...	X	...	...	...	...	...
columbiae McDunnough	...	...	...	X	X	...	...	...	X	...	...	...
croesus McDunnough	...	...	...	...	...	...	X	...	...	...	...	...
midas McDunnough	...	...	...	...	...	...	X	...	...	...	...	...
Species?	...	...	...	...	...	X	...	...	...	...	...	...

Siphlonisca Needham (figs. 79, 242, 356)

Nymphal Characteristics. Body length: 19–20 mm. Tubercles present on mesosternum and metasternum of thorax. Posterolateral projections on abdominal segments 5–9 greatly expanded. Gills single, subcordate, with abundant tracheation. Terminal filament slightly shorter than cerci. Illustration: Edmunds, Allen, and Peters, 1963, pl. V.

Nymphal Habitat. Small temporary pools resulting from stream overflow.

Nymphal Habits. Not described.

Life History. Little known. The earliest observation of nymphs was made on May 25, and emergence was noted on the same day; by May 28 the pool was completely dry, but all nymphs had emerged. Typically, the nymph crawls out of the water on a rush stem to a height of three or four inches; emergence occurs at all hours of the day but is most common in late morning. Emergence dates extend to June 18.

Adult Characteristics. Wing length: 17 mm. Eyes of male large, contiguous dorsally, not divided. Midventral tubercles present on thoracic mesosternum and metasternum. Forelegs of male shorter than body; basal tarsal segment longer than second segment; blunt spinelike processes present on fore tibiae and fore tarsal segments. Claws similar. Hind wings with costal projection rounded. Abdominal segments 5–9 with wide lateral expansions. Male genitalia illustrated in fig. 356.

Mating Flights. Reported as occurring high in the air.

Taxonomy. Needham, 1908. *New York State Mus. Bull.* 134:72. Type species: *S. aerodromia* Needham; type locality: Sacandaga Park, New York. The single species is known only from New York.

Distribution. Reported to be Holarctic, but records from Japan need confirmation of generic identity. Known in Nearctic only from upper New York State. Since so little is known of this genus and its distribution is so restricted we are including some previously unpublished details on the type locality area. Apparently C. P. Alexander and G. C. Crampton are the only entomologists who have seen *Siphlonisca aerodromia* alive. Alexander (personal communication) has generously provided us with important data about this area in New York.

About 1907 while collecting there I noted this striking insect flying high above the river, easily told even at a distance by the greatly widened abdomen. In those days the river at Northville and Sacandaga Park had a major island, called Sport Island, connected with the mainland at the Park by a long wooden bridge. It was while standing on this bridge over the river that the mayflies, craneflies, etc., were easily observed and many could be collected. The first one taken was sent to the then New York State Entomologist, Dr. E. P. Felt and he sent it to Needham who described it in the 1908 paper. After this I collected a few further specimens that went into the Cornell collection, etc. I entered Cornell in 1909 and was with Needham thereafter for the next eight years. He always was much interested in *Siphlonisca* and in 1914 (while I was an Instructor in the Entomology Department) allowed me to get away early to go to this part of the Sacandaga to see if we could locate the nymphs of the mayfly. At that time the river was lowering from the high spring floods and there were numerous small shallow pools along the low sandy margins and in these there was a great concentration of aquatic life, including several species of mayflies, with *Siphlonisca*. These pools gradually dried up and the unemerged life was destroyed.

When Alexander returned to the east in 1922, he joined the staff of what is now the University of Massachusetts. One of his colleagues there was G. C. Crampton, who also had a special interest in *Siphlonisca*, apparently because the large abdominal projections of *Siphlonisca* seemed reminiscent of some Carboniferous insects. Alexander collected four males on June 18, 1924, and another male on June 4, 1925, all of which he gave to Crampton. (The specimens are now at the University of Utah.as a gift from the Crampton Collection at the Museum of Comparative Zoology.)

Alexander recalls that Crampton was not successful in finding *Siphlonisca* until the 1930s, when the Sacandaga Reservoir flooded the whole northeastern part of Fulton County. "After this . . . there remained of the original Sacandaga River only the parts in southern Hamilton Co., and especially the main stream from Northville almost to Wells, where the two main branches unite to form the Sacandaga River. From what Crampton told me I believe that he found his . . . specimens in this section of the stream south of Wells."

Alexander believes that Crampton's specimens were taken in the vicinity of the Sacandaga public campground. "The section of the river below Wells has not been changed since Crampton's days and I believe that the species still holds out in the twenty-mile stretch from the Campground south to Northville. Furthermore, it does not seem logical that the species should have any such limited range as this and there seems to be no reason why it should not occur in others of the major streams of the southern Adirondack Mountain area, as the Hudson River itself, north of Glens Falls, etc. I have never been at these sites in the proper season, that is, probably early June, but others may be able to relocate the mayfly."

Siphlonurus Eaton (figs. 61, 62, 67, 72, 75, 221, 240, 245, 392)

Nymphal Characteristics. Body length: 9–20 mm. Claws moderately long and slender. Posterolateral projections on apical abdominal segments moderately to strongly developed. Gills double on abdominal segments 1 and 2 only, or all seven pairs double; gills large, with abundant tracheation, apical margins retuse near middle. Illustration: Fig. 392.

Nymphal Habitat. The nymphs are usually found in shallow, quiet pools or ponds along the edges of streams where they climb among the vegetation or move over the bottom. They are sometimes found in shallow pools of seepage water, pools on rock ledges, and in shallow pools fed intermittently with fresh water. Some species occur along the margins of lakes. *Siphlonurus autumnalis* is found in very shallow water beneath boulders on the edges of large rivers. Young nymphs of some species may be found in flowing water; mature nymphs migrate to quiet waters which may become isolated pools.

Nymphal Habits. The nymphs of all species swim well but spend most of their time on silty bottoms; when disturbed they swim rapidly to another place, then come to rest again on the bottom. When the nymphs are at rest, their large gills are in constant motion. They swim by rapidly vibrating the caudal filaments. The legs are held posterolaterally when the nymphs swim but are extended when they rest on the bottom. All or most species are omnivorous, opportunistically feeding on plant and insect remains and capturing soft-bodied bottom-feeding insects, especially tube-dwelling chironomids and mosquitoes. *S. occidentalis* appears to select its food by size; the nymphs pick up particles of the proper size and rotate them with the palpi, at times swallowing even particles of gravel of the acceptable size.

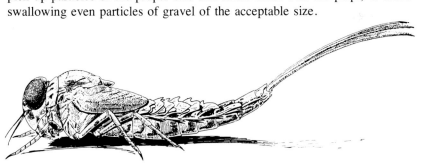

Fig. 392. *Siphlonurus alternatus*, nymph. (From Burks, 1953.)

Life History. Emergence of adults occurs most commonly from April through July, but in some species as late as October. Emergence is in morning or early afternoon. The nymphs crawl entirely out of the water to emerge. The subimago stage lasts from twenty-six to fifty-three hours. The length of nymphal life has not been determined for North American species, but for most species there is probably only one generation each year. In Europe four species overwinter as eggs, and the nymphs develop in two or three months during the summer. A fifth species, *S. aestivalis*, has two generations during the summer months under favorable conditions. For most species the winter months appear to be spent in the egg stage.

Adult Characteristics. Wing length: 9–20 mm. Usually strikingly marked; head and thorax usually mostly dark brown, abdomen with conspicuous pattern of contrasting light and dark areas producing annulate appearance. Eyes of male large, usually contiguous dorsally, not divided. Forelegs of male as long as or slightly longer than body; basal tarsal segment at least three-fourths length of second segment, sometimes subequal to it; first and second tarsal segments with blunt spinelike processes on margin; sharp spines on margin of fore tibiae. Claws similar. Costal angulation of hind wings broadly rounded. Male genitalia illustrated in fig. 240.

Mating Flights. *Siphlonurus* adults have been observed swarming in midafternoon over the shoal margin of an Adirondack lake and also in late afternoon flying over a streamside highway. Many species have been observed flying in late afternoon or early evening, and some of these also swarm in midmorning. In Idaho *S. occidentalis* has been observed swarming ten to thirty feet over a wet highway just at dark; mating pairs from the swarms settled on the top of a parked automobile. *Siphlonurus* has a long ascending and descending flight with no swaying from side to side. *Siphlonurus quebecensis* and also *S. rapidus* have been observed mating and laying eggs in Michigan, generally over water but sometimes over nearby roads whose dark surfaces often solicit the same behavioral response as do rivers for many mayflies. The females of some species hover a few feet above the water when ovipositing, occasionally dropping to the water surface for as long as thirty seconds, then rising and repeating the performance. Females of *Siphlonurus occidentalis* have been observed in early morning, rapidly flying a few inches above stream riffles and invariably depositing their eggs in the riffles.

Taxonomy. Eaton, 1868. *Entomol. Mon. Mag.* 5:89. Type species: *S. flavidus* (Pictet); type locality: San Ildefonso, Spain. The species of adults should be identifiable with relative ease. The nymphs can often be recognized by their distinctive ventral abdominal color patterns. The nymphs of a number of species have not been described.

Distribution. Holarctic. In the Nearctic, reaching southern limits in the mountains of California, Arizona, and New Mexico, in the Ozarks in Missouri, and in the Appalachians south to South Carolina.

	United States						Canada			Mexico		Central America
Species	SE	NE	C	SW	NW	A	E	C	W	N	S	
alternatus (Say)	...	X	X	...	...	...	X	X	X	...	...	...
autumnalis McDunnough	...	...	...	...	X	...	...	...	X	...	...	...
barbaroides McDunnough	...	X	...	...	...	...	X	...	...	...	...	...
barbarus McDunnough	...	X	...	...	...	...	X	...	...	...	...	...
columbianus McDunnough	...	...	X	X	X	...	...	...	X	...	...	...
decorus Traver	X	...	...	...	...	...	...	...	...	...	...	...
luridipennis (Burmeister)	X	...	...	...	...	...	...	...	...	...	...	...
marginatus Traver	X	...	...	...	...	...	...	...	...	...	...	...
marshalli Traver	...	...	X	...	...	...	...	...	...	...	...	...

Species	United States						Canada			Mexico		Central America
	SE	NE	C	SW	NW	A	E	C	W	N	S	
mirus Eaton	X	X	...	...	...	...	...	...	...	...	...	...
occidentalis Eaton	...	...	X	X	X	...	...	...	X	...	...	...
phyllis McDunnough	...	...	...	...	...	...	X	...	X	...	...	...
quebecensis (Provancher)	X	X	X	...	...	...	X	...	...	...	...	...
rapidus McDunnough	...	X	X	...	...	...	X	...	...	...	...	...
securifer McDunnough	...	X	...	...	...	...	X	...	...	...	...	...
spectabilis Traver	...	...	...	X	...	...	...	...	...	...	...	...
typicus Eaton	...	X	X	...	...	...	X	...	...	...	...	...
Species?	...	...	...	...	...	X	...	...	...	...	...	...

Subfamily Acanthametropodinae

This subfamily is Holarctic in distribution, being known from China, Siberia, Wyoming, Utah, Illinois, Georgia, and South Carolina. It is one of the least known subfamilies of mayflies and is extremely rare in collections. Two genera occur in North America.

Nymphal Characteristics. Body somewhat flattened. Antennal length about twice width of head. Mouthparts of carnivorous type, mandibles without molar area. Legs all directed posteriorly; tibiae strongly bowed. Claws long and slender. Median tubercles present on prothoracic and mesothoracic sterna, present or absent on metasterna.

Adult Characteristics. Hind wings very large, being half or more as long as forewings. Styliger plate of male deeply excavated on the mesal posterior margin. Terminal filament not vestigial.

Acanthametropus Tshernova (fig. 69)

Nymphal Characteristics. Body length: 20 mm or more. Head with one pair of short submedian tubercles between the antennal bases. Maxillary palpi absent. Projecting tubercles present on lateral margins of pronotum and mesonotum; a hooklike tubercle on each thoracic sternum. A median hooklike tubercle on each abdominal tergum. Gills with inner margins deeply dissected. Illustration: Burks, 1953, fig. 312 (as *Metreturus*).

Nymphal Habitat. Rapid, shallow rivers with sand and rock bottoms. The nymphs are clearly adapted for living in sand. They have been collected in only three large rivers in the world — the Rock River in Illinois, Three Runs, a tributary of the Savannah River in South Carolina, and the Amur River in Siberia.

Nymphal Habits. Unknown.
Life History. Unknown.
Adult Characteristics. Unknown.
Mating Flights. Unknown.
Taxonomy. Tshernova, 1948. *Dokl. Akad. Nauk SSSR*, n.s., 60(8):1453. Type species: *A. nikolskyi* Tshernova; type locality: Transbaikal, Amur River basin, Siberia. It is possible that the generic name *Acanthametropus*, based on nymphs, is synonymous with the genus *Siphluriscus*, which is known only as adults from China. See discussion of taxonomy under *Analetris*. The only Nearctic species is *A. pecatonica* (Burks) (described as *Metreturus*). The nymphs from South Carolina and Illinois differ in the shapes of the thoracic tubercles, but it is impossible to judge whether the observed differences are individual variations or specific differences; part of the differences appear to result from the age of the nymphs.
Distribution. Holarctic. In Nearctic, known only from Illinois and South Carolina.

Analetris Edmunds (figs. 70, 71, 223, 357, 393)

Nymphal Characteristics. Body length: 15 mm. Head without frontal or lateral tubercles. Maxillary palpi three-segmented. Pronotum and mesonotum without lateral tubercles; hooklike tubercles present on prosternum and mesosternum. Abdominal terga without median tubercles. Gills with margins entire, and a recurved posterior (ventral) flap with two portions; trachea pale. Illustration: Fig. 393.
Nymphal Habitat. Moderate to large warm rivers on sandy bottoms with some silt. Known only from the Green River in Utah and Wyoming, the Blacks Fork River in Wyoming, and the South Saskatchewan River in Saskatchewan, Canada. The Green River habitats are now submerged in Flaming Gorge Reservoir, and the Blacks Fork River is threatened with pollution by agricultural and domestic wastes, salt water, and oil spillage from oil drilling. A new reservoir constructed on the headwaters may change the water temperature or the flow pattern in the stream below it. Portions of the Saskatchewan River are also polluted.
Nymphal Habits. The nymphs lie on the sand with the legs directed posteriorly; they are strong swimmers when disturbed. They prey on chironomid larvae.
Life History. Unknown. Nymphs have been collected from June through September. The life cycle appears to take one year or less; a subimago male was reared in the laboratory on August 9 from a nymph collected on August 3.
Adult Characteristics. (Based on male subimago.) Wing length: 12 mm. Eyes of male nearly contiguous dorsally; eyes bicolored. Hind wings without

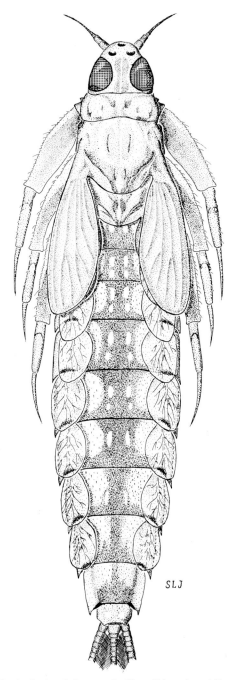

Fig. 393. *Analetris eximia*, nymph. (From Edmunds and Koss, 1972.)

costal projection. Claws dissimilar. Fore tibiae less than half as long as femora; tarsi about 1½ times as long as tibiae; tarsi in order of decreasing length: 1, 2 = 5, 3 = 4. Male genitalia as in fig. 357. Caudal filaments about as long as abdomen; terminal filament almost as long as cerci.

Mating Flights. Unknown.

Taxonomy. Edmunds and Koss, 1972. *Pan-Pac. Entomol.* 48:138. Type species: *A. eximia* Edmunds; type locality: Blacks Fork River at Interstate 80, west of Green River (city), Wyoming. There is a single species. The only member of the Siphlonuridae with which it can possibly be confused is the undescribed adult of *Acanthametropus*. There is a possibility that the nymphs named *Acanthametropus* are the same or a close ally of the adults named *Siphluriscus*, known only from adult forms in China. If so, *Siphluriscus* has the terminal filament only 2 or 3 mm long rather than subequal to the cerci as in *Analetris*. In all other Holarctic Siphlonuridae the adults have only a vestigial terminal filament.

Distribution. Nearctic. Known only from a short section of the Blacks Fork River in Sweetwater County, Wyoming, and from the South Saskatchewan River in Saskatchewan, Canada. Formerly found in the Green River in Sweetwater County, Wyoming, and Daggett County, Utah.

Subfamily Isonychiinae

This subfamily contains a single widespread genus, *Isonychia*, which is distributed in the Holarctic, Oriental, Nearctic, and Neotropical realms. The genus is remarkable in that it retains some very primitive features (for instance, its tracheal system), although it has rather specialized feeding modifications. The proper classification of the genus is difficult. It has many characters suggesting close affinity with the Oligoneuriidae, and there is strong evidence that *Isonychia* has pregroup, group, postgroup relationships with these two families as follows: "proto-Siphlonuridae" gave rise to proto-*Isonychia*, which in turn gave rise to Oligoneuriidae. Its intermediate position between the Siphlonuridae and the Oligoneuriidae and near the base of the Coloburiscinae of the Siphlonuridae has led many mayfly taxonomists to place *Isonychia* as the family Isonychiidae, which may either include or exclude the Coloburiscinae. See *Isonychia* for characteristics and details on distribution in North and Central America.

Isonychia Eaton (figs. 45, 216, 222, 358, 394)

Nymphal Characteristics. Body length: 9–17 mm. Head with median frontal ridge below middle ocellus; mouthparts hairy. Antennal length up to 2½ times width of head. Gill tufts present at bases of maxillae. Fore tibiae with a conspicuous apical spine; forelegs with a double row of long setae on inner surface of femora and tibiae. Gill tufts present at bases of fore coxae. Claws stout and denticulate on inner margin. Gills consist of dorsal lamellae and fibrilliform portion. Illustration: Fig. 394.

Nymphal Habitat. Vigorous swimmers, the nymphs are found in rapidly flowing waters of creeks and rivers, occurring in tangles of vegetation and debris anchored in the stream, especially on branches and collections of leaves caught in areas of swift flow. They may be found concentrated in large numbers where branches trail in flowing water. They are also found on rocks or on flat rock ledges where there is considerable disturbance of the water as it pours across the surface of the rocks. Negatively phototropic.

Nymphal Habits. Very strong swimmers. The nymphs face the current with the forelegs held in front of the mouth, the long hairs overlapping, and filter food from the flowing water. From time to time they graze on materials caught on the hairs. Leonard and Leonard (1962) report that diatoms and algae comprise much of the diet, but some species also feed on the larvae of midges and blackflies and on the nymphs of smaller mayflies.

Life History. In warmer parts of the southeastern coastal plain, emergence may occur throughout the year; in colder areas emergence is limited to the warmer months. The length of time required for nymphal development has not been determined, but it probably requires considerably less than a year. The nymph crawls a few inches above the water surface onto a rock, a stick, or any other protruding object for the subimago to emerge. The length of the sub-imago stage varies between twenty-two and thirty-one hours. Emergence occurs in the late afternoon or shortly after dark, with an occasional subimago appearing in early morning.

Adult Characteristics. Wing length: 9–16 mm. Usually reddish to purplish brown; male and female coloration very similar. Femora and tibiae of forelegs dark; fore tarsi, middle and hind legs, and tails pale cream to white in many species. Eyes of male large, contiguous dorsally. Remnants of gill tufts persist at sides of vestigial maxillae and at bases of fore coxae. Spinous median projections present at anterior margins of mesosternum and metasternum and near middle of mesosternum. Forelegs of male somewhat shorter than body; fore tarsi approximately as long as fore tibiae. Costal angulation of hind wings obtuse. Male genitalia illustrated in fig. 358. Two caudal filaments present, sometimes with vestige of terminal filament.

Mating Flights. Most species swarm at twilight, and large numbers may congregate over a stream. Other species fly in midafternoon or midmorning. Some species fly twenty feet or more above the water, with mating couples or individuals drifting down to stream level and then back into the swarm. *Isonychia sicca campestris* flies five to ten feet above the water. The females

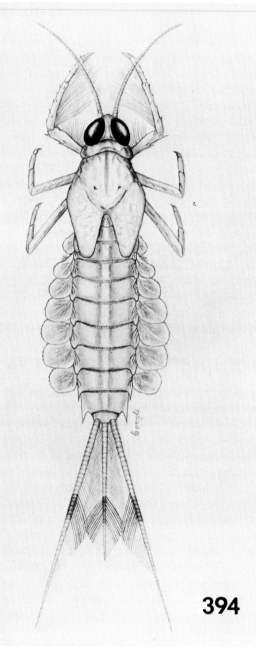

394

Fig. 394. *Isonychia pictipes*, nymph. (From Berner, 1950.)

oviposit while flying downstream, then reverse their flight and touch the abdomen to the water surface to release the nearly spherical egg mass. Many species release the egg mass from a height of several feet.

Taxonomy. Eaton, 1871. *Trans. Entomol. Soc. London*, 1871, p. 134. Type species: *I. sicca manca* Eaton; type locality: Texas. Specific identification is not difficult with most adults, but nymphs cannot be identified with certainty because some characteristics used in the keys are known to vary with age. Locally, mature nymphs should be identifiable once the adults have been identified and associated with the nymphs.

Distribution. Holarctic, Oriental, and Neotropical. In America the genus is widespread but generally more abundant in the eastern United States and Canada. The genus occurs in Mexico and as far south as the mountains of Honduras.

Species	United States						Canada			Mexico		Central America
	SE	NE	C	SW	NW	A	E	C	W	N	S	
annulata Traver	X											
arida (Say)			X									
aurea Traver	X											
bicolor (Walker)	X	X	X				X					
christina Traver	X	X					X					
circe Traver	X											
diversa Traver	X											
fattigi Traver	X											
georgiae McDunnough	X											
harperi Traver		X	X									
intermedia (Eaton)				X						X		
matilda Traver		X										
notata Traver	X											
obscura Traver	X											
pacoleta Traver	X											
pictipes Traver	X											
rufa McDunnough	X	X	X									
sadleri Traver	X	X	X				X					
sayi Burks			X									
serrata Traver	X											
sicca campestris McDunnough				X	X			X	X			
sicca manca Eaton			X	X						X		
sicca sicca (Walsh)	X		X				X					
similis Traver	X											
thalia Traver	X											
tusculanensis Berner	X											
velma Needham				X	X							
Species?											X	X

Family METRETOPODIDAE

This small family of Heptagenioidea appears to be a derivative of the proto-siphlonurine Siphlonuridae rather than a relative of the genus *Ametropus*. The two included genera are very closely allied, although the adults differ quite markedly. The nymphs of *Metretopus* are more delicate, softer, and more slender. The two genera are not easily separated as nymphs, partly because the nymphs of all of the *Siphloplecton* are not known and partly because the known species have diverse gill types. However, the keys should allow for discrimination of the two genera. The adults are readily distinguished by the venation of the forewings. The distribution is Holarctic.

Nymphal Characteristics. Body streamlined, generally minnowlike. Tarsi of all legs longer than tibiae. Claws of forelegs bifid. Claws of second and third pairs of legs slender, longer than tibiae. Caudal filaments with long setae on both sides of terminal filament and on mesal sides of cerci.

Adult Characteristics. Forewings with one or two pairs of intercalaries between CuA and CuP; if two pairs present, vein MP_2 strongly divergent from MP_1 at base. Stigmatic crossveins anastomosed or tending to anastomose. Fore tarsi of male about three times as long as tibiae; hind tarsi four-segmented. Claws dissimilar on all tarsi. Two caudal filaments.

Metretopus Eaton (figs. 130, 297, 372)

Nymphal Characteristics. Body length: 9–10 mm. Abdomen slender, tergum 9 about 1¼ times as wide as long. Gills on abdominal segments 1–7 single, similar in form. Illustration: None, except Fig. 130.

Nymphal Habitat. The nymphs occur mostly in slow-flowing streams on vegetation or on the bottom near the shore.

Nymphal Habits. Not described.

Life History. Adults have been taken in late June in Michigan and in late July and mid-August in Canada.

Adult Characteristics. Wing length: 9 mm. Brownish. Compound eyes of male large, nearly contiguous dorsally. Wings hyaline; vein MP_2 of forewings only slightly divergent from MP_1 at base; one pair of cubital intercalary veins present; crossveins in stigmatic area somewhat anastomosed. Male genitalia illustrated in fig. 372.

Mating Flights. Not described.

Taxonomy. Eaton, 1901. *Entomol. Mon. Mag.*, 2d Ser., 12:253. Type species: *M. borealis* Eaton (as *norvegicus* Eaton); type locality; Al, Norway. Only one Holarctic species has been reported from the New World.

Distribution. Holarctic. The only Nearctic species, *M. borealis*, is distributed completely across Canada and south into Michigan and Maine.

Siphloplecton Clemens (figs. 58, 133, 134, 236, 298, 373, 395)

Nymphal Characteristics. Body length: 9–16 mm. Abdomen rather robust, tergum 9 almost twice as wide as long. Gills on abdominal segments 1–2 or 1–3 with posterior (ventral) recurved flap or flaps; gills single on segments 3–7 or 4–7. Illustration: Fig. 395.

Nymphal Habitat. In slow-flowing large streams the nymphs occur close to shore among vegetation in water from three inches to two feet deep. In Michigan *S. basale* has been reported from fairly deep water with a strong current in medium-sized to large rivers. One nymph of *Siphloplecton* sp. was collected by deep dredging from Lake Superior by the United States Fish and Wildlife Service. The nymphs also occur at the shores of lakes.

Nymphal Habits. The nymphs swim very actively and vigorously in short spurts.

Life History. Unknown. Development takes less than one year. Subimagoes taken in Georgia required approximately forty-eight hours to molt. In the southeast adult specimens have been taken from late March through mid-April; farther north, adults have been taken in June. In Michigan emergence appears to peak during the last two weeks of May, but adults have been collected as late as July 15.

Adult Characteristics. Wing length: 9–16 mm. Chiefly brown with paler markings. Compound eyes of male contiguous dorsally, not divided. Wings usually marked with dark clouds at base and bulla and with costal and discal crossveins infuscated (wings hyaline in one species); vein MP_2 of forewings strongly divergent from MP_1 at base; two pairs of cubital intercalary veins present; crossveins of stigmatic area anastomosed, forming two rows of cells. Male genitalia illustrated in fig. 373.

Mating Flights. Swarms of *Siphloplecton* have been reported only for *S. basale* in Michigan (Leonard and Leonard, 1962). The species is easily recognized, even in flight, by its large size, spotted wings, and vigorous flight. "Males in mating fly back and forth at a height of 20–30 feet above the water, with little rising and falling. Females entering the swarm are quickly seized, and the pair at once rise in a towering flight which carries them over the top of stream-side trees. Females were observed to deposit their eggs in a sort of dive-bombing flight, the eggs being released from a height of 4 to 8 feet above the water. Sometimes, however, the females settled to the surface, where they drifted with the current for up to half a minute."

Taxonomy. Clemens, 1915. *Can. Entomol*. 47:258. Type species: *S. basale* (Walker); type locality: Lake Winnipeg, Manitoba. The adult males are readily identifiable to species, but no keys are available to identify nymphs to species. The adult color pattern, which can generally be seen in mature

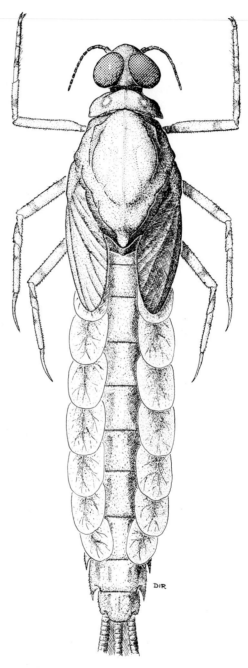

Fig. 395. *Siphloplecton* sp., nymph.

nymphs, serves as an aid in identification. A revisionary study has been initiated by Lewis Berner.

Distribution. Nearctic, most common in the eastern United States and Canada, extending west to central Alberta and south to Illinois and Florida.

Species	United States						Canada			Mexico		Central America
	SE	NE	C	SW	NW	A	E	C	W	N	S	
basale (Walker)	X	X	X	...	...	...	X	X	X	...	...	...
costalense												
Spieth	X	...	...	...	...	...	...	...	...	...	...	...
interlineatum												
(Walsh)	...	...	X	...	...	...	...	X	...	...	...	...
signatum Traver	X	...	...	...	...	...	...	...	...	...	...	...
speciosum Traver	X	...	...	...	...	...	...	...	...	...	...	...

Family SIPHLAENIGMATIDAE

This family is recognized for a single New Zealand species, *Siphlaenigma janae* Penniket. It is clearly intermediate between the Baetidae and Siphlonuridae. The nymphal behavior is quite similar to that of other stream Baetidae, but the abdominal ganglia and Malpighian tubules are siphlonurine. The placement of a single species in a separate family is a questionable practice, but it must be weighed against the difficulty of defining the Baetidae or Siphlonuridae if *Siphlaenigma* is included in either family. To be consistent with classificatory practice in the mayflies, the taxon might better be placed as a subfamily of the Siphlonuridae.

Nymphal characteristics. Body streamlined, minnowlike. Head hypognathous. Glossae and paraglossae of labium somewhat long and narrow but broader than those of Baetidae. Gills on abdominal segments 1–7 simple and lanceolate to elongate and oval in shape. Abdominal segments without posterolateral projections. Three caudal filaments about equal in length.

Adult Characteristics. Eyes of male not divided. Forewings with MA forming a regular fork; IMP and MP_2 detached basally from MP. Interspaces between veins at outer margins of forewings without short intercalaries. Hind wings small but with abundant venation. Penes of male somewhat reduced but distinct and bilobed. Two caudal filaments only.

Family AMETROPODIDAE

The Ametropodidae appear to be Heptagenioidea that diverged from the ancestral stock early and now constitute an isolated group. The family is recognized for the single genus *Ametropus*. The various other genera that have been placed in the family are now assigned to Pseudironinae of the Heptageniidae, Acanthametropodinae, and Siphlonurinae of the Siphlonuridae or to the Metretopodidae. Long claws are primarily an adaptation for sand dwelling, and this adaptation seems to influence the proportions of other parts of the leg in adults and nymphs. These characters have been used to justify the placement of genera of diverse origins in the family.

Ametropus is known from North America, Siberia, European Russia, and Poland and formerly from Holland. See *Ametropus* for characteristics and distribution in North America.

Ametropus Albarda (figs. 57, 224, 374, 396)

Nymphal Characteristics. Body length: 14–18 mm. Body flattened. Frontal margin of head reduced almost to bases of antennae so that mouthparts are almost entirely exposed. Legs slender; forelegs very short with spinous pads attached at base of coxae. Claws of forelegs slender and slightly curved, bearing four or five long spines; similar spines on tarsi and tibiae. Claws of middle and hind legs long and straight, each subequal in length to tarsi and about twice length of tibiae, without spines or denticles. Flattened lateral projections on abdominal segments 2–9, with short posterolateral projections on segments 8–9. Gills single, obovate, margined with long setae. Three caudal filaments; terminal filament with heavy fringe of setae on both sides, cerci heavily fringed on inner margins. Illustration: Fig. 396.

Nymphal Habitat. The nymphs bury themselves in firm, slightly silty sand in large rivers with a relatively strong current. The sand must be firm, smooth, and clean; in most rivers the nymphs are found only in a very small percentage of the sandy bottom. In the Blacks Fork River from Granger, Wyoming, to Interstate 80, localities have been found with fifteen or more nymphs per square meter. Nymphs have been taken in New Mexico from a warm river at an altitude of 5,400 feet; they occur in similar rivers in Utah, Colorado, Wyoming, Montana, Oregon, Washington, Alberta, and Saskatchewan, although the northern records are from lower altitudes.

Nymphal Habits. Good swimmers. The nymphs swim with the middle and hind legs trailing to the sides and the forelegs held in front of the head; they settle into the sand until only the dorsum, the gills, and the eyes protrude. The forelegs are used constantly to groom the frons and the antennae. The coxal pads characteristic of this genus are pressed to the sand and probably function

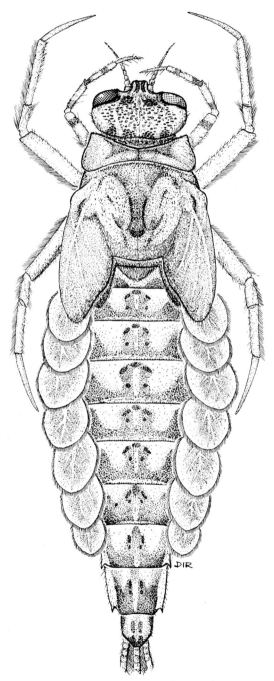

Fig. 396. *Ametropus albrighti*, nymph.

to help the nymph maintain its position. The nymphs appear to feed on the microbiota, principally algae, that are found interstitially in the sand. Nymphs reared in the laboratory continued to grow from September to May on the food in the sand brought from the natural habitat. The aquarium was kept lighted continually to allow the growth of algae.

Life History. Adults have been collected from May through July. *A. albrighti* requires almost one year for development. Young nymphs appear shortly after the adults emerge and become mature the next year. The nymphs are highly variable in size, indicating an emergence period of a month or more.

Adult Characteristics. Wing length: 13–15 mm. Eyes of male with larger facets in upper portion. Fore tarsi of male very long, about five times as long as short tibiae; hind tarsi four-segmented; basal tarsal segment and tibia fused; claws dissimilar on all tarsi. Forewings with two pairs of cubital intercalaries, anterior pair longer and attached to vein CuP basally; vein A_1 attached to hind margin by series of veinlets. Hind wings with acute costal projection. Male genitalia illustrated in fig. 374. Three caudal filaments.

Mating Flights. Adults believed to be *Ametropus* because of their size have been observed about thirty feet above a large river in midafternoon. Mature nymphs of *Ametropus* were in the river at the time.

Taxonomy. Albarda, 1878. *Entomol. Mon. Mag*. 15:129. Type species: *A. fragilis* Albarda; type locality: Arnhem, Holland. Allen and Edmunds are in the process of preparing keys and figures to the nymphs and adults of the North American species.

Distribution. Holarctic. In the Nearctic, *A. albrighti* Traver is known only from tributaries of the Colorado River, and *A. neavei* McDunnough occurs in Alberta and Saskatchewan. A third undescribed species occurs in Washington, Oregon, and Montana.

Family BAETIDAE

This family is a member of the Heptagenioidea. The Baetidae almost certainly have been derived from proto-*Metamonius*-complex siphlonurine Siphlonuridae of Chile, Argentina, Australia, and New Zealand. The New Zealand genus *Siphlaenigma* (Siphlaenigmatidae) is almost perfectly intermediate between the Siphlonuridae and the Baetidae and must have been derived from an early pre-baetid. In North America the Siphlonurinae most closely related and most similar to the Baetidae, especially as nymphs, are members of the genus *Ameletus*.

The family is widespread, being found on all continents and on many islands. It is absent from New Zealand, although its closest relative, *Siphlaenigma*, occurs there. At extremely high northern latitudes and in high altitude streams of North America and Asia, *Baetis* is the only mayfly genus present.

Generic limits in the family are generally vague and poorly defined. Many more species must be reared to solve the generic problems. The generic assignment of several United States and Canadian species is doubtful, and such problems are much more severe for Mexican and Central American species.

It is certain that the loss of the hind wings in the Baetidae has occurred repeatedly and that the present generic classification, which separates species into one genus if hind wings are present and into another if hind wings are absent, is artificial. This subject is also discussed in the taxonomy of *Cloeon*. Early mayfly workers placed all species lacking hind wings in either *Cloeon* (if the intercalaries of the forewings occurred singly) or *Pseudocloeon* (if the intercalaries of the forewings occurred in pairs). Thus, the Central American and South American species assigned to *Pseudocloeon* may be either *Pseudocloeon*, *Baetodes*, *Cloeodes* (described from Puerto Rico), *Paracloeodes*, *Apobaetis*, or an ally of *Baetis* from New Mexico which also has no hind wing pads.

The nymphs which we have keyed out as *Pseudocloeon* in couplet 42 form a group distinct from *Baetis* because of their broad and flattened form. This group is probably a derivative of the *lapponicus* group of *Baetis*, in which the nymphal form is broader than in most species of *Baetis*, the terminal filament is very short, and the adults have hind wings which are minute in the females and small and without a costal projection in the males. Because we suspect that the loss of hind wings has occurred in other *Baetis* groups, we have taken the precaution of keying out other nymphs without hind wing pads as *Pseudocloeon* because their adults would be placed as *Pseudocloeon*. The genus *Pseudocloeon* was established for a species from Java, and it is not certain that the American species are congeneric with the Javanese species.

The absurdity of the present classification does not escape us. We are plagued further by the knowledge that a series of adults collected in Idaho, which appears to represent one species, has hind wings in the males but not in the females. There is also a species of Baetidae in Argentina in which only the males have hind wings.

We do not propose to change the obviously invalid generic classification of the Baetidae at this time because we believe that a useful classification can be achieved only after considerable study of reared material from throughout the world. The generic classification in the Baetidae is almost certain to be based largely on the nymphs; new adult characters must be sought for assigning the species to genus. The primary center of evolution of the Baetidae was South America. There is evidence that several lines may have spread from South America to Africa during the period of geological history when the two continents were connected or were less distantly separated. Africa is clearly a secondary center of the early evolution of the Baetidae, but there is no evidence to suggest that the Baetidae evolved before temperate Africa was separated in the south from the Gondwanaland supercontinent (see Edmunds, 1972). The recognition of South America as a major center of evolution for the Baetidae has important implications for the taxonomy of this difficult family.

The genera *Baetodes*, *Dactylobaetis*, and *Cloeodes* are clearly austral, and we have only slightly less confidence that *Apobaetis* and *Paracloeodes* are austral, although neither is known from south of the United States. Only one additional generic name, *Camelobaetidius*, has been applied to South American Baetidae. In our nymphal key *Camelobaetidius* will key out to *Dactylobaetis*; it differs from *Dactylobaetis* in having only a minute terminal filament and in having a ventral tubercle at the base of each fore femur. *Camelobaetidius* may be encountered in Panama.

In the warm waters of the United States, especially in the southern part, and in Mexico and Central America there is an increase in the number of problem species of Baetidae. Several of the unusual nymphs which seem to be *Baetis* or *Pseudocloeon* allies are not keyed out because they lack names and their affinities are uncertain. Nymphs from New Mexico collected by R. W. Koss have extremely large gills, each being much larger than the segment from which it arises, but the claws and caudal filaments are *Baetis*-like. These will key to *Baetis* in couplet 39, but they have no hind wing pads and so the adults would presumably key to *Pseudocloeon*. One Central American type has large mature nymphs (length approximately 10 mm), whose mandibular incisors are fused into a flat scraper. These will key to *Baetis*, and the adults apparently will have the characters of *Baetis*. Some nymphs have claws with slender separate denticles unlike those of most *Baetis* or with even finer denticles; at least some of these are *Baetis*.

The authors have seen other unusual Baetidae from North America and especially from South America that would cause difficulties in the keys and in classification. It is possible that some of these occur in Central America. We have already mentioned the Idaho specimens whose nymphs should have wing pads in the male only.

Nymphal Characteristics. Body streamlined, minnowlike. Head hypognathous. Antennae long, usually two or more times as long as width of head.

Labium with glossae and paraglossae long and narrow. Gills on abdominal segments 1–7, 1–5, or 2–7. Abdominal segments without posterolateral projections except in *Callibaetis*, *Cloeon*, and some *Centroptilum* species, in which they may be moderately developed.

Adult Characteristics. Eyes of male divided, upper portion turbinate. Forewings with veins IMA, MA$_2$, IMP, and MP$_2$ detached basally. Interspaces between veins at outer margin of forewing with short single or paired intercalaries. Hind wings reduced, with or without veins, with one to three longitudinal veins, or with hind wings absent. Two caudal filaments only. Penes of male membranous and retractable, extrusions illustrated in fig. 397.

Apobaetis Day (figs. 101, 108, 256, 259, 262, 359)

Nymphal Characteristics. Body length: 4 mm. Labrum with straight, unmodified apical margin. Apex of distal segment of labial palpi squarely truncate and set with stiff spines, a distinct angular notch on inner apical margin. Claws of all legs very long and slender, fully as long as tarsi, almost straight, and without denticles. Hind wing pads absent. Gills ovate and single, main tracheal trunk pigmented, lateral branches not pigmented. Three caudal filaments; terminal filament about as long as cerci, heavily margined with fine setae on each side; cerci heavily margined with setae on mesal side. Illustration: Not illustrated in full; parts in Day, 1955, figs. 1–6.

Nymphal Habitat. The only species was collected from a warm river which had been dredged, diverted, dammed, and polluted (from irrigation runoff, crop dusting, sewage, and industry). The water temperatures were as high as 28° C (82° F). The nymphs were found in small groups over widely scattered areas, occurring in fine sand at depths of eighteen to twenty-four inches in rather rapidly flowing water.

Nymphal Habits. Not described.

Life History. The length of time required for development is unknown. Adults have been reared in August.

Adult Characteristics. Wing length: 4 mm. Turbinate eyes of male very large, high, and slender. Metascutellum in lateral view prominent but not flattened and projecting; metathoracic postnotum plus epimeron deeply emarginate medially (figs. 259, 262). Forewings with marginal intercalaries in pairs; intercalaries about as long as width of space between the members of a pair; hind margin of basal half of forewing rather evenly rounded. Hind wings absent. Male genitalia illustrated in fig. 359.

Mating Flights. Not described.

Taxonomy. Day, 1955. *Pan-Pac. Entomol.* 31:127. Type species: *A. indeprensus* Day; type locality: Tuolumne River, Stanislaus County, California. Only a single species of this genus is known. It is not unlikely that other species will be discovered, especially in Mexico or in the southwestern United States.

Distribution. Nearctic; the single species is known from only one locality in California.

Baetis Leach (figs. 60, 68, 95–99, 109, 110, 112, 113, 233, 247, 249, 251, 252, 360, 397, 398)

Nymphal Characteristics. Body length: 3–12 mm. Head relatively short and high; antennae inserted at about midpoint on height of head. Labrum with narrow but distinct median notch on apical margin. Distal segment of labial palpi rounded; second segment usually enlarged apically on the mesal surface (figs. 96, 98, 99). Claws with several to many denticles, the apical denticle largest. Hind wing pads present; may be minute. Gills present on abdominal segments 1–7 or 2–7, single on all segments; in most species gills obovate, but in some species gills of segments 6 and 7 narrow and lanceolate. In most species, three caudal filaments; in a few species only two caudal filaments with vestigial terminal filament. In most species with three caudal filaments, terminal filament shorter and thinner than cerci; long setae present on both sides of terminal filament and on mesal side of cerci. Illustration: Berner, 1950, pl. XXII.

Nymphal Habitat. Usually in shallow flowing water. Most commonly in riffles, on or under stones and rocks, among debris, or on or among vegetation along banks of creeks or rivers. Depending on the species, the nymphs may be found in relatively quiet and slow-flowing waters or in torrential streams. At very high altitudes in the Rocky Mountains (also in the Himalaya in Asia) or in cold streams at high latitudes (for instance, in Alaska or on Southampton Island at the north end of Hudson Bay), the two-tailed species of *Baetis* are the only mayflies present. A few species live near lakeshores where there is continuous wave action. Some species are tolerant of high temperatures, others are restricted to cold mountain or northern streams. Some species make up a substantial proportion of the insects found in stream drift.

Nymphal Habits. The nymphs swim well; when they stop, they seize some point of attachment with their claws and raise the abdomen well above the support. The gills are held close to and over the abdomen or laterally. The nymphs may dart in short spurts from one spot to another by rapidly flipping the caudal filaments up and down. In those species with three caudal filaments the setae on the filaments overlap to form an effective organ for rapid swimming. The nymphs cling closely to resting places, heading upstream with the abdomen swinging from side to side in the current. Positive rheotropic orientation is maintained regardless of whether the nymphs are on the upper side or the underside of the support. They crawl slowly over the surface of leaves or

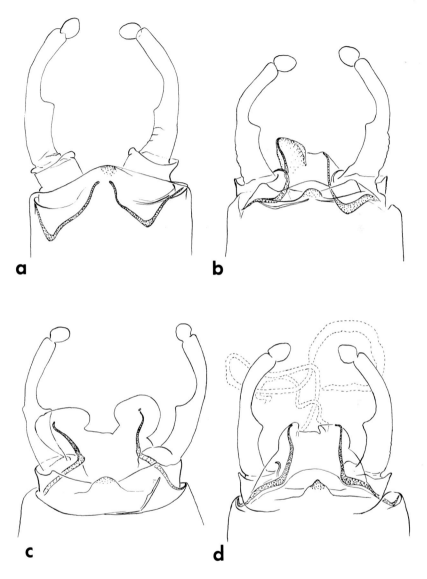

Fig. 397. Progressive stages in eversion of the membranous
baetid penes of *Baetis tricolor* Tshernova. (Redrawn
from photographs in Keffermüller, 1972.)

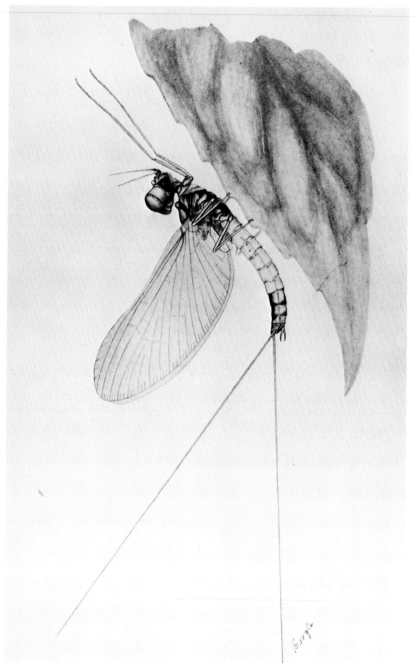

Fig. 398. *Baetis spinosus*, male. (From Berner, 1950.)

rocks feeding on small particles. They appear to feed entirely on plants and detritus.

Life History. In Florida adults emerge throughout the year. Farther north the season of most species is much more restricted, and the principal emergence occurs during the summer months. Adults of *B. tricaudatus*, however, emerge throughout the year. The time required for nymphs to mature varies with the area and the species. In warm streams there may be more than two generations per year or per summer. *B. parvus* in Oregon has only one generation per year. The life histories of American species are probably as diverse as those of European species. In Czechoslovakia one species requires eleven months at high altitudes; other species have a one-year cycle, but the length of nymphal life is variable. Some species have two generations per year, a summer brood developing in about three months and an overwintering brood taking a longer time. One species overwinters in the egg stage but has two generations in the summer. Females of several species have been observed ovipositing; they alight on a partially submerged object and crawl into the water, where they attach the eggs to the object in rows. The wings of the submerged females collapse onto or alongside the abdomen. The eggs of one species hatch in the laboratory in twenty-eight days, but the hatching period varies considerably in European species and presumably also in North American species. One species is reported to pass through twenty-seven instars. Emergence occurs during daylight hours, the time of emergence varying with the species and the season. The subimago stage lasts from seven to twelve hours. When ready to emerge, the nymph floats at the surface of the water and the subimago bursts free almost immediately (in five or ten seconds). In the laboratory turbulent water is necessary for successful emergence in some species.

Adult Characteristics. Wing length: 3–11 mm. Forelegs with tarsi from slightly shorter than tibiae to longer than tibiae; basal segment of tarsi very short. Forewings with marginal intercalaries in pairs. Hind wings much reduced, sometimes difficult to see; long and narrow; with or without an acute costal projection; with one, two, or three longitudinal veins. Abdomen of male with segments 2–6 usually hyaline, white, or light brown, other segments dark; abdomen of female unicolorous, usually brown or reddish brown. Male genitalia illustrated in figs. 360, 397.

Mating Flights. In most species flights occur in late morning or early afternoon. The flights may consist of few to many individuals, and they generally take place over water or over a stream bank at a height of six to fifteen feet. *B. tricaudatus* in Utah has been found swarming in openings in coniferous forests in late afternoon or early evening, and in Washington it has been observed swarming in the lee of an isolated hilltop conifer about one mile from the nearest stream.

Taxonomy. Leach, 1815. *Brewster's Edinburgh Encyclopedia* 9:137. Type species: *B. fuscatus* (Linnaeus); type locality: (Neotype) Sweden. The adults of this genus are particularly difficult to identify to species because many are very similar and there may be a number of synonymies. Hind wing venation, so much used in differentiating species, is highly variable. Further, summer broods of emerging adults are often much smaller and paler than the spring broods of the same species. The nymphs are perhaps even more difficult to identify than the adults. There are many minute but good characters to facilitate the differentiation of species — particularly the mouthparts, the texture of the abdominal segments, the relative length of the caudal filaments (a characteristic much less constant than indicated in the existing keys), and the leg and claw structure. In North and Central America this important genus will almost certainly have to be studied in great detail, and keys will have to be constructed on a regional basis. The species groups recognized by Müller-Liebenau (1970) for European species of *Baetis* appear to be applicable to many of the North American species and will serve as a valuable reference for grouping the American species.

Distribution. Cosmopolitan except in New Zealand and on oceanic islands. In the Nearctic, extremely widespread from Alaska into Central America.

	United States						Canada			Mexico		Central America
Species	SE	NE	C	SW	NW	A	E	C	W	N	S	
adonis Traver	...	...	...	X	...	...	...	...	...	...	...	...
akataleptos McDunnough	...	...	...	...	...	...	...	...	X	...	...	...
alius Day	...	...	...	X	...	...	...	...	...	...	...	...
amplus (Traver)	X	...	...	...	...	...	...	...	...	...	...	...
anachris Burks	...	...	X	...	...	...	...	...	...	...	...	...
australis Traver	X	...	...	...	...	...	...	...	...	...	...	...
baeticatus Burks	...	...	X	...	...	...	...	...	...	...	...	...
bicaudatus Dodds	...	...	...	X	X	...	...	...	...	...	...	...
brunneicolor McDunnough	...	...	X	...	...	...	X	...	...	...	...	...
bundyae Lehmkuhl	...	...	...	...	...	...	...	...	X	...	...	...
caurinus Edmunds & Allen ...	...	...	...	...	X	...	...	...	...	...	...	...
cleptis Burks	...	...	X	...	...	...	...	...	...	...	...	...
devinctus Traver	...	...	...	X	...	...	...	...	...	...	...	...
diablus Day	...	...	...	X	...	...	...	...	...	...	...	...
eatoni Kimmins	...	...	...	...	...	...	...	...	...	X	...	...
elachistus Burks	...	...	X	...	...	...	...	...	...	...	...	...
endymion Traver	...	...	X	...	...	...	...	...	...	...	...	...
ephippiatus Traver	X	...	...	...	...	...	...	...	...	...	...	...

Species	United States						Canada			Mexico		Central America
	SE	NE	C	SW	NW	A	E	C	W	N	S	
erebus Traver	...	...	...	X	...	...	...	...	...	...	...	...
flavistriga McDunnough	X	X	...	...	...	...	X	...	...	...	...	...
foemina McDunnough	...	...	...	...	...	...	X	...	...	...	...	...
frivolus McDunnough	...	...	X	...	...	...	X	...	...	...	...	...
frondalis McDunnough	X	...	X				X	...	...	...	...	...
hageni Eaton	X	X	X	...	...	...						...
harti McDunnough	...	...	X	...								...
hiemalis Leonard	...	...	X	...				...		...		...
hudsonicus Ide	...	...	...	...	...	...	...	X	...	...		...
incertans McDunnough	X	...	...	...	...	...	X	...	...	...	...	...
insignificans McDunnough	...	...	...	X	X	...	...	...	X	...	...	...
intercalaris McDunnough	X	X	X	...	...	...	X	...	...	...	...	...
intermedius Dodds	...	...	...	X	X	...	...					...
jesmondensis McDunnough	...	...	...	...	X	...	...	...	X	...	...	...
lapponicus (Bengtsson)	...	...	...	...	...	...	X	...	...	...	...	...
leechi Day	...	...	...	X	...	...	...	...	...	...	...	...
levitans McDunnough	X	X	X	...	...	...	X	...	...	...	...	...
macdunnoughi Ide	...	...	...	...			X	...	...	...		...
moffati Dodds	...	...	...	X								...
nanus McDunnough ...	...	...	X	...	...	...	X	...	...	...	...	...
ochris Burks	...	...	X	...								...
palisadi Mayo	...	...	...	X								...
pallidulus McDunnough	X	...	X	...	...	...	X	...	...	...	...	...
parallelus Banks	...	...	...	X	...							...
parvus Dodds	...	...	...	X	X	...	X	...	X	...	...	...
persecutus McDunnough	...	...	...	...	...	...	...	...	X	...	...	...
phoebus McDunnough	...	...	X	...	...	...	X	...	...	...	...	...
phyllis Burks	...	...	X	...								...
piscatoris Traver	...	...	...	X								...
pluto McDunnough	...	X	X	...	...	...	X	...	...	...	...	...
posticatus (Say)	...	X	...	...								...
propinquus (Walsh)	X	...	X	X	X	...	X	X	...	...	...	...
pygmaeus (Hagen)	...	X	X	...	...	...	X	...	...	...	...	...
quebecensis McDunnough	X	X	X				X	...	...	...	...	...

Species	United States						Canada			Mexico		Central America
	SE	NE	C	SW	NW	A	E	C	W	N	S	
quilleri Dodds	...	...	...	X	X	...	...	...	...	...	...	...
rusticans McDunnough	...	X	...	...	...	...	X	...	...	...	...	...
salvini Eaton	...	...	...	...	...	...	...	...	...	...	...	X
sinuosus Navas	...	...	...	...	...	...	...	...	...	...	...	X
spiethi Berner	X	...	...	...	...	...	...	...	...	...	...	...
spinosus McDunnough	X	X	X	...	...	...	X	X	...	...	...	...
sulfurosus Day	...	...	...	X	...	...	...	...	...	...	...	...
thermophilos McDunnough	...	...	...	X	X	...	...	...	...	...	...	...
tricaudatus Dodds	...	...	X	X	X	...	...	...	...	X	...	...
vagans McDunnough	...	X	X	...	...	...	X	...	...	...	...	...

Baetodes Needham and Murphy (figs. 80, 81, 253, 260, 264, 361)

Nymphal Characteristics. Body length: 3–8 mm. Legs exceptionally long, held out from body. Claws short and stout, bearing a series of denticles. Hind wing pads absent. Ventrally directed gills present on abdominal segments 1–5 only; with a row of median raised projections or dense tufts of hair on the middle abdominal terga. Two caudal filaments; terminal filament reduced to stub; cerci bare or with a few inconspicuous setae. Illustration: Traver, 1944, fig. 11.

Nymphal Habitat. In rapids of medium-sized streams the nymphs cling to the upper surface of rocks or to vegetation. The nymphs are well camouflaged. In Arizona they have been observed to occupy the same section of stones on which black fly larvae were concentrated.

Nymphal Habits. The nymphs crawl very slowly. They swim poorly, quite unlike most Baetidae.

Life History. Emergence occurs in May and June in Arizona and apparently throughout the year in Brazil.

Adult Characteristics. Wing length: 4 mm. Turbinate eyes of male large, cylindrical, and erect. Metascutellum in lateral view small and inconspicuous; metathoracic postnotum and epimeron deeply emarginate medially (figs. 260, 264). Forelegs as long as body; tibiae 1½ to more than 2 times length of tarsi; basal fore tarsal segment equal in length to second and third segments combined. Forewings with crossveins in disc of wing very fine, inconspicuous; marginal intercalaries in pairs; intercalaries about as long as width of space between members of a pair; basal half of hind margin of forewing partly subparallel to costal margin; cubitoanal margin somewhat angular. Hind wings absent. Short median tubercles may be present on anterior abdominal terga on or near posterior margin. Male genitalia illustrated in fig. 361.

Mating Flights. Unknown.

Taxonomy. Needham and Murphy, 1924. *Bull. Lloyd Lib.* 24, Entomol. Ser. 4:55. Type species: *B. serratus* Needham and Murphy; type locality: Tijuca, Rio de Janeiro, Brazil. There are many species in this genus, but not all have been named. Adults cannot be identified at present. Most of the species have been described as nymphs recently by Koss, Mayo, or Cohen and Allen. Although keys have not been published, the species have many distinctive characters and should be identifiable.

Distribution. Neotropical and Nearctic. The greatest species diversity is in the Neotropical region, with some species extending into the Nearctic as far north as Texas and Arizona.

Species	United States						Canada			Mexico		Central America
	SE	NE	C	SW	NW	A	E	C	W	N	S	
adustus Cohen & Allen	...	...	...	...	...	...	...	...	...	...	X	...
arizonensis Koss	...	...	...	X	...	...	...	...	...	...	...	...
bellus Mayo	...	...	...	...	...	...	...	...	...	...	X	...
caritus Cohen & Allen	...	...	...	...	...	...	...	...	...	...	X	X
deficiens Cohen & Allen	...	...	...	...	...	...	...	...	...	...	X	X
edmundsi Koss	...	...	...	X	...	...	...	...	...	...	...	...
fortinensis Mayo	...	...	...	...	...	...	...	...	...	...	X	
furvus Mayo	...	...	...	...	...	...	...	...	...	...	X	
fuscipes Cohen & Allen	...	...	...	...	...	...	...	...	...	...	X	X
inermis Cohen & Allen	...	...	...	...	...	...	...	...	...	X	X	...
longus Mayo	...	...	...	...	...	...	...	...	...	...	X	
noventus Cohen & Allen	...	...	...	...	...	...	...	...	...	...	...	X
obesus Mayo	...	...	...	...	...	...	...	...	...	...	X	...
pallidus Cohen & Allen	...	...	...	...	...	...	...	...	...	...	X	X
pictus Cohen & Allen	...	...	...	...	...	...	...	...	...	...	X	...
tritus Cohen & Allen	...	...	...	...	...	...	...	...	...	X	X	...
veracrusensis Mayo	...	...	...	...	...	...	...	...	...	...	X	...

Callibaetis Eaton (figs. 64, 65, 93, 94, 230, 243, 362, 399)

Nymphal Characteristics. Body length: 6–10 mm. Apex of distal segment of labial palpi rounded or somewhat pointed. Claws long and slender, with double row of minute, slender denticles. Hind wing pads present. Gills on abdominal segments 1–4 and usually 5–7 with ventral recurved flap or flaps (in some species flap is small, in others gills on anterior segments appear to be double or triple); all gills with densely pinnate tracheation. Three caudal filaments about equal in length and thickness; terminal filament fringed on both sides with long setae; cerci fringed on mesal side only; the apical portion of the caudal filaments is bare in some species. Illustration: Fig. 399.

Nymphal Habitat. In still water such as permanent ponds, roadside ditches, and margins of lakes or in transient pools. The nymphs show very wide limits of tolerance, occurring in great abundance in areas where the water is choked with vegetation, but also occurring in areas where the vegetation is very sparse. The temperature of the water in which nymphs occur may sometimes rise to 32° C (90° F) in the southern portion of Florida, but some species in the northern part of the range and in the mountains develop at temperatures as low as 4° C (40° F). The nymphs appear to be very tolerant of either acid or alkaline waters, and one species has been reported from brackish water. Some species are also found among the vegetation in backwaters of streams or along the margins where there is little flow. Nymphs of *Callibaetis* have also been found in the water in the leaf axils of bromeliads in Mexico.

Nymphal Habits. Among the most graceful of all mayfly immatures. With the body arched, the insect hangs from a plant stem or sits on the bottom where it becomes virtually invisible except when it moves. When disturbed, it darts away swiftly by rapid flicks of the caudal filaments; when removed from the water, it flips about much like a small minnow. The nymphs are herbivorous, feeding on diatoms and other algae.

Life History. Certain species require six weeks for development. The eggs are retained, developing within the body of the female for a period of five or six days. Each female produces 450 to 500 eggs. The female drops onto the surface of the water and releases the eggs, which hatch immediately; the young seek and cling to vegetation. Nine to sixteen instars are required for development. The adults usually emerge in afternoon or early evening. The subimaginal molt takes place seven to nine hours after emergence. Females have been kept alive as long as thirteen days in the laboratory. The males die soon after mating. Adult males were seen mating with subimago females on one occasion.

Adult Characteristics. Wing length: 6–12 mm. Upper portion of eyes of male with short stalk, expanded dorsally. Bodies of both sexes with numerous reddish brown or brown spots set in small depressions. Sexual dimorphism pronounced with pigmentation in female more intense than in male. Forewings with marginal intercalaries either in pairs or single; some species with relatively few crossveins, forming a single irregular row across the wing; others with numerous crossveins, forming two or more irregular rows across the wing. Forewings of male with basal costal crossveins weak or lacking,

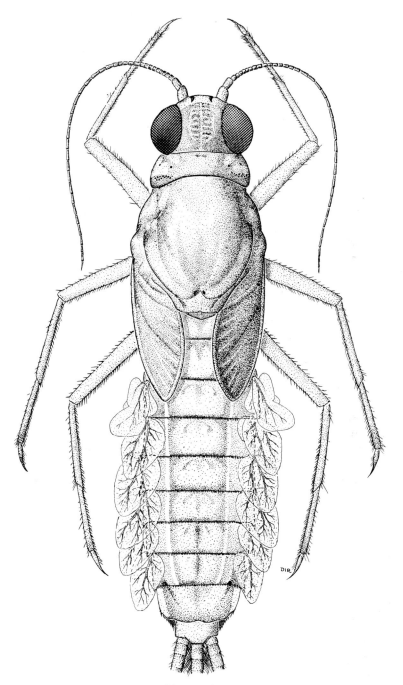

Fig. 399. *Callibaetis coloradensis*, nymph.

well developed in female; forewings of male usually not pigmented, those of female brown at least in the costal and subcostal interspaces. Hind wings with costal projection large and obtuse; numerous crossveins. Male genitalia illustrated in fig. 362.

Mating Flights. In full sunlight from early morning to midafternoon over open meadows near water. The swarms very in height from four to twenty-five feet, rising and falling from a few inches to two feet. Three species in Utah all swarm over areas without vegetation. After mating the female retreats to a resting place until the eggs are ready to hatch.

Taxonomy. Eaton, 1881. *Entomol. Mon. Mag.* 17:196. Type species: *C. pictus* (Eaton); type locality: Texas. In many parts of North America, especially in the southwest and in Mexico, the genus is particularly abundant. In most of the United States those species with a relatively small number of crossveins in the forewing may be identified to species with a reasonable degree of confidence. The species having more numerous crossveins are much more difficult to identify and are more poorly known. There are several reasons for experiencing difficulty in identifying adults of *Callibaetis*. The marked sexual dimorphism makes it difficult to associate males and females of the same species correctly unless they are reared or are taken *in copula*. The extent of individual variation in many species appears to be immense; for example, in *C. nigritus* Banks the first adults that emerge in early spring are twice as large as those that emerge in late summer. The females of the spring brood have an almost solid black vitta across the front of the wings, but females of the late summer brood may have only a few scattered dark specks along the wing margin. The male genitalia seem to provide only a small range of characters. The nymphs show a variety of character differences, but as yet the only workable keys are for relatively small regions.

Distribution. Neotropical and Nearctic. Very widespread in the Nearctic north to Alaska.

Species	United States						Canada			Mexico		Central America
	SE	NE	C	SW	NW	A	E	C	W	N	S	
americanus Banks	...	X	...	X	X	...	X	...	...	...	...	...
brevicostatus Daggy	...	...	X	...	...	...	...	X	...	...	...	...
californicus Banks	...	...	...	X	...	...	...	...	...	...	...	...
carolus Traver	...	...	...	X	...	...	...	...	...	...	...	...
centralis Peters	...	...	X	...	...	...	...	...	...	...	...	...
coloradensis Banks	...	...	X	X	X	...	...	...	X	...	...	...
doddsi Traver	...	...	...	X	X	...	...	...	...	...	...	...
evergreenensis Thew	...	...	...	X	...	...	...	...	...	...	...	...

Species	United States						Canada			Mexico		Central America
	SE	NE	C	SW	NW	A	E	C	W	N	S	
ferrugineus (Walsh)		X	X									
floridanus Banks	X											
fluctuans (Walsh)	X	X	X				X					
hebes Upholt				X								
montanus Eaton			X	X	X							X
nigritus Banks				X	X							
pacificus Seemann			X	X	X							
pallidus Banks				X								
paulinus (Nevas)												X
pictus (Eaton			X	X								X
pretiosus Banks	X	X	X				X					
semicostatus Banks				X				X				
signatus Banks				X								
skokianus Needham		X	X				X					
traverae Upholt				X								
undatus (Pictet)											X	X
Species?						X				X		

Centroptilum Eaton (figs. 84–87, 90, 92, 102, 103, 231, 232, 363, 400)

Nymphal Characteristics. Body length: 4–8 mm. Distal segment of labial palpi dilated and truncate apically. Claws one-third to more than one-half as long as tarsi, with or without small denticles. Gills usually asymmetrical with median side broader; gills on anterior segments (usually on segment 1 or segments 1 and 2, in some species as far back as segment 6) may have a recurved, dorsal flap; lateral tracheal branches usually much better developed on mesal side. Three caudal filaments; terminal filament as long as cerci, with long setae on both sides; cerci with long setae on mesal side; caudal filaments with narrow dark bands at apex of every third to fifth segment in basal two-thirds. Illustration: Fig. 400.

Nymphal Habitat. Some species live among lily pads and pond weeds in quiet portions of streams; other species occur in moderately swift water on upper surfaces of stones or on vegetation. Some nymphs may be found on the sand along the margins of a large river. The nymphs are tolerant of slow-flowing water, and they may be found along the edges of lakes where there is wave action; very few species live in ponds. One species from Florida is confined to streams which are alkaline (pH between 7.3 and 8.0).

Nymphal Habits. The nymphs have tails that are depressed slightly at the tips; they swim easily by rapidly flicking the abdomen. The nymphs are very active, and when one is taken out of water, it flips like a small minnow. The nymphs feed on plant material.

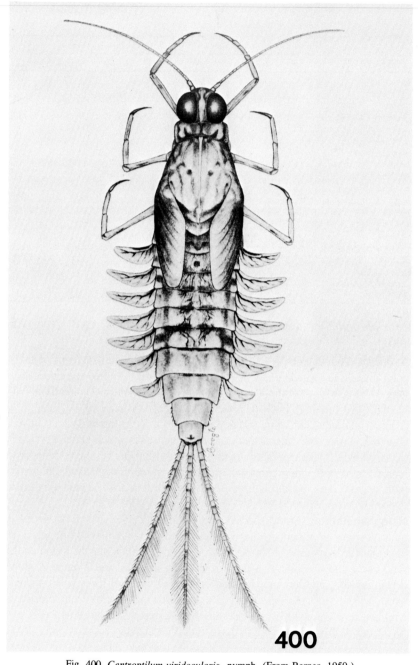

Fig. 400. *Centroptilum viridocularis*, nymph. (From Berner, 1950.)

Life History. The length of the nymphal stage has not been determined, but it probably requires six to nine months. Emergence takes place generally in late afternoon. The subimagoes molt within ten to twelve hours after emergence. Some evidence indicates that the time interval of later nymphal instars is between three and five days.

Adult Characteristics. Wing length: 4–8 mm. Male often with head, thorax, and basal and terminal abdominal segments dark brown, middle abdominal segments pale but sometimes marked dorsally with red or reddish brown; female generally light colored, abdominal terga often marked with black tracheal lines. Turbinate eyes of male with short stalk; upper surface usually oval in shape. Forelegs of male with tarsi slightly longer than tibiae. Forewings with marginal intercalaries single. Hind wings small, relatively long and slender with hooked costal projection; two longitudinal veins in most species. Male genitalia illustrated in fig. 363.

Mating Flights. Adults have been seen in the evening swarming two to six feet above highways and areas cleared of vegetation and also in midmorning swarming above small streams. A female has been seen ovipositing at sunset in October in Florida.

Taxonomy. Eaton, 1869. *Entomol. Mon. Mag.* 6:132. Type species: *C. luteolum* (Muller); type locality: Europe. In general the males of *Centroptilum* should be identifiable to species by use of coloration, genitalia, and form of hind wings. The nymphs are not yet well enough known to provide a workable key to species.

Distribution. Holarctic, Ethiopian, Oriental, and Australian. In the Nearctic, widespread south to northern Mexico.

	United States						Canada			Mexico		Central
Species	SE	NE	C	SW	NW	A	E	C	W	N	S	America
album												
McDunnough.......	...	...	X	...	...	...	X	...	...	...	...	...
asperatum Traver	...	...	...	X	...	...	...	...	...	...	...	...
bellum												
McDunnough.......	...	...	X	...	...	...	X	...	...	...	...	...
bifurcatum												
McDunnough.......	...	...	...	...	X	...	...	...	X	...	...	...
caliginosum												
McDunnough.......	...	...	...	...	...	...	X	...	...	...	...	...
conturbatum												
McDunnough.......	...	...	...	X	X	...	...	...	X	...	...	...
convexum Ide	...	...	X	X	...	...	X	...	...	...	...	...
elsa Traver	...	...	...	...	X	...	...	...	...	...	...	...
fragile												
McDunnough.......	X	X	...	...	...	...	X	...	...	...	...	...
hobbsi Berner	X	...	...	...	...	...	...	...	...	...	...	...

Species	United States						Canada			Mexico		Central America
	SE	NE	C	SW	NW	A	E	C	W	N	S	
infrequens McDunnough	...	...	...	...	...	...	...	X	...	...	...	...
intermediale McDunnough	X	...	...	...	...	...	X	...	...	...	...	...
oreophilum Edmunds	...	...	...	X	X	...	...	...	...	...	...	...
ozburni McDunnough	...	X	...	...	...	...	X	...	...	...	...	...
quaesitum McDunnough	...	...	X	...	...	...	...	...	X	...	...	...
rivulare Traver	...	...	X	...	...	...	...	...	...	...	...	...
rufostrigatum McDunnough	...	...	X	...	...	...	X	X	...	...	...	...
selanderorum Edmunds	...	...	...	X	X	...	...	...	...	...	...	...
semirufum McDunnough	...	...	...	...	...	...	X	...	...	...	...	...
similie McDunnough	...	X	...	...	...	...	X	...	...	...	...	...
venosum Traver	...	...	...	X	...	...	...	...	...	...	...	...
victoriae McDunnough	...	...	...	...	...	...	X	...	...	...	...	...
viridocularis Berner	X	...	...	...	...	...	...	...	...	...	...	...
walshi McDunnough	...	...	X	...	...	...	...	...	...	...	...	...
Species?	...	...	...	...	...	X	...	...	...	X	...	...

Cloeon Leach (figs. 66, 88, 89, 91, 104–106, 246, 364)

Nymphal Characteristics. Body length: 4–9 mm. Distal segment of labial palpi truncate at apex. Claws usually one-half or more as long as tarsi; with or without small denticles. Hind wing pads absent. Gills on abdominal segments 1–7 oval, either double, single, and broad, or with recurved dorsal flap on anterior segments (usually segment 1 or segments 1–2, in some species as far back as segment 6); tracheae branched more or less palmately or with branches better developed on mesal side. Three caudal filaments approximately equal in length and thickness; setae present on both sides of terminal filament and on mesal side of cerci; caudal filaments with a dark band at apex of every fourth to fifth segment. Illustration: Eaton, 1883–88, pl. 47.

Nymphal Habitat. In slow-flowing streams or in backwaters or quiet areas along the margins of more rapid streams. The nymphs live among vegetation, sometimes occurring in large numbers. They have also been taken at the edge of some large lakes and have been found in weedy areas at the edge of ponds in the northern part of the range.

Nymphal Habits. The nymphs climb about on vegetation. They are apparently herbivorous. Some species are known to eat detritus, diatoms, and other small algae.

Life History. In Florida no definite seasonal period for emergence. Farther north, in colder water, emergence is restricted to the warmer months. When ready to emerge, the nymph rises to the surface, and the subimago escapes from the nymphal skin almost immediately. Observations have indicated that upon emergence female subimagoes may be immediately seized by imago males. The subimaginal period lasts about six or seven hours. In Europe viviparous species have been reported, and it is likely that this method of reproduction also occurs in New World forms as well.

Adult Characteristics. Wing length: 3–9 mm. Head, thorax, and basal and terminal abdominal segments brown in males of many species; middle abdominal segments pale or pale with reddish markings. Females paler than males; yellowish to pale orange or reddish brown. Turbinate eyes of male with short stalks. Forelegs of male slightly shorter than body; tarsi subequal to tibiae. Forewings with relatively few crossveins; marginal intercalaries single; hind wings absent. Male genitalia illustrated in fig. 364.

Mating Flights. Unknown.

Taxonomy. Leach, 1815. *Brewster's Edinburgh Encyclopedia* 9:137. Type species: *C. dipterum* (Linnaeus); type locality: Europe. Generally, the adults of this genus should be identifiable to species. In some species individual variants may have either brown or reddish pigmentation. The nymphs of certain species may be identified with some degree of reliability.

The presently recognized genera *Centroptilum* and *Cloeon* are, at least in the Holarctic region, a natural group. The separation of the two genera is based on the presence of a hind wing in *Centroptilum* and its absence in *Cloeon*. We no longer recognize *Neocloeon* Traver as separate from *Cloeon*, an opinion expressed earlier by Ide (1937) and concurred in by Traver (personal communication), and we question the validity of separating *Centroptilum* and *Cloeon*.

It appears that the primitive nymph of *Centroptilum* had simple oval gills (fig. 90) and rather short claws (fig. 103). One part of the genus became modified by elongation of the claws and the addition of an anterior (dorsal) flap on the gills of segment 1 or segments 1–2, probably through selection for life in slow-moving water. Further modification resulted in loss of the hind wings, thus giving rise to *Cloeon*. Further changes within *Cloeon* resulted in increased gill surface for life in ponds, as in species of *Cloeon* with larger recurved flaps or with additional numbers of gills with such flaps. Two species in North America, *C. alamance* and *C. triangulifer*, have elongate claws and greatly enlarged gills without recurved flaps. It is possible that these species have lost the recurved gill flaps rather than having lost the hind wings independently of other *Cloeon*, but judgment should be deferred until more is known of character clusters in relation to gills and claw lengths.

Cloeon may be a monophyletic specialized offshoot of *Centroptilum*, or it may represent several derivations from *Centroptilum*. There seems to be little point in lumping *Cloeon* and *Centroptilum* into a single genus until the relationships of the various species are better known. However, it should be pointed out that Crass (1947) has placed an African species without hind wings in *Centroptilum* and that we have seen other species without hind wings closely related to another African species group assigned to *Centroptilum*. Hence on a world basis some forms assignable to *Cloeon* seem to represent additional origins from *Centroptilum*.

Distribution. Ethiopian, Oriental, Australian (but not New Zealand), oceanic islands (Samoa, Yap), Neotropical, and Holarctic. In the Nearctic, most abundant in northeastern United States and eastern Canada south to Idaho, Colorado, Illinois, and Florida. The Neotropical species were assigned to *Cloeon* by earlier authors whose concept of the genus was broader than the current one, and the occurrence of *Cloeon* in the Neotropical region is doubtful.

Species	United States						Canada			Mexico		Central America
	SE	NE	C	SW	NW	A	E	C	W	N	S	
alamance (Traver)	X	...	X	...	...	...	...	...	...	...	...	...
dipterum (Linnaeus)	...	...	X	...	...	...	...	...	...	...	...	...
implicatum McDunnough	...	...	...	...	X	...	...	...	X	...	...	...
inanum McDunnough	...	...	...	...	...	...	...	...	X	...	...	...
ingens McDunnough	...	X	...	X	...	...	X	...	X	...	...	...
insignificans McDunnough	...	...	...	...	...	...	X	...	...	...	...	...
mendax (Walsh)	...	X	X	...	...	...	X	...	...	...	...	...
minor McDunnough	...	...	...	...	...	...	X	...	...	...	...	...
rubropictum McDunnough	X	X	X	...	...	...	X	...	...	...	...	...
simplex McDunnough	...	...	...	...	...	...	...	...	X	...	...	...
triangulifer McDunnough	...	...	...	...	...	...	X	...	...	...	...	...
vicinum (Hagen)	...	X	...	...	...	...	...	...	...	...	...	...

Dactylobaetis Traver and Edmunds (figs. 82, 83, 248, 365)

Nymphal Characteristics. Body length: 3.5–10.0 mm. Maxillae with twisted palpi. Short, threadlike gills present or absent near base of forelegs;

legs usually short and stout, forelegs usually longer than other two pairs; tarsi distinctly bowed; claws flattened, with from five to forty or more apical denticles. Simple obovate gills on abdominal segments 1–7 with first and seventh pairs smaller than other pairs. Three caudal filaments; terminal filament from three-fourths length of cerci to almost same length as cerci. Illustrations: Not illustrated in full; parts in Traver and Edmunds, 1968, numerous illustrations.

Nymphal Habitat. Habitats diverse, but the nymphs are known to occur in rocky, slowly to swiftly flowing streams that may be silty at least during part of the year. In the western United States all streams from which nymphs were collected were silty and reached relatively high summer temperatures with daytime water samples ranging from 15.5° C (60° F) to 29.5° C (85° F). Nymphs of *D. warreni* were found in shallow, fast riffles less than four inches deep over pebbles (one-half inch to one inch in diameter) mixed with sand. Nymphs believed to be *D. cepheus* were taken from gravel at depths to one foot. Nymphs of *D. mexicanus* were collected from the exposed roots of plants in flowing water at the banks of streams. The northward extension of the range of *Dactylobaetis* may be limited to silted streams because other streams may not reach suitably high temperatures. As with many other southern genera whose most northerly occurrence is limited to silty streams, these mayflies are found in clear streams in Mexico, Central America, and South America.

Nymphal Habits. Not recorded; probably similar to *Baetis*.

Life History. Unknown. Adults have been collected or reared in California during June, July, and August. Adults are known from Idaho in August, from northern Mexico in December, and from Central America in August, November, and December.

Adult Characteristics. Wing length: 4.0–7.5 mm. Turbinate eyes of male with short to medium stalks. Forewings with marginal intercalaries in pairs; stigmatic crossveins simple, slanting, few in number. Hind wings with two veins only; costal projection acute and broadly based; anterior margin beyond apex of costal projection more or less undulate. Males of some species with head, thorax, and terminal abdominal segments reddish brown; middle abdominal segments paler. Male genitalia illustrated in fig. 365.

Mating Flights. Adults have been observed in Idaho swarming at the edge of a road late in the evening toward the end of August.

Taxonomy. Traver and Edmunds, 1968. *Pac. Insects* 10:629–677. Type species: *D. warreni* Traver and Edmunds; type locality: Tuolumne River, Stanislaus County, California. Most of the species are known only as nymphs and are keyed by Traver and Edmunds (1968). Almost certainly there are a number of undescribed species, especially in Mexico and Central America.

Distribution. Neotropical and Nearctic. The greatest species diversity occurs in the Neotropical region, with some species extending into the Nearctic north to Oregon, Saskatchewan, and Oklahoma.

Species	United States						Canada			Mexico		Central America
	SE	NE	C	SW	NW	A	E	C	W	N	S	
arriaga Traver & Edmunds	...	...	...	...	...	...	...	...	...	...	X	...
cepheus Traver & Edmunds	...	...	...	...	X	...	...	...	...	...	...	...
chiapas Traver & Edmunds	...	...	...	...	...	...	...	...	...	...	X	...
jensensi Traver & Edmunds	...	...	...	...	...	...	...	...	...	...	X	...
mexicanus Traver & Edmunds	...	...	...	...	...	...	...	...	...	X	...	...
musseri Traver & Edmunds	...	...	...	...	...	...	...	...	...	...	X	X
warreni Traver & Edmunds	...	...	...	X	...	...	...	...	...	...	...	...
zenobia Traver & Edmunds	...	...	...	...	...	...	...	...	...	...	...	X
Species?	...	...	X	...	...	...	...	...	...	...	...	...

Heterocloeon McDunnough (fig. 250)

Nymphal Characteristics. Body length: 6–8 mm. Head relatively short and high; antennae inserted at about mid-point of height of head. Labrum with rather broad median notch on apical margin. Distal segment of labial palpi rounded, the second segment not enlarged apically on the mesal surface. Claws with two dissimilar rows of denticles, the apical denticle of the one row largest. Hind wing pads present but minute. Short threadlike thoracic gills ventrally near base of forelegs. Gills present on abdominal segments 1–7, single on all segments. Two long caudal filaments, the terminal filament reduced to a single segment; basal three-fourths of cerci with long setae on the mesal margin. Illustrations: Parts of all species in Müller-Liebenau, 1974.

Nymphal Habitat. The nymphs live in small to large fast-flowing streams. One of the species was found in crevices of rocks on which *Podostemon ceratophyllum* Michaux was growing, and it is possible that this is the typical habitat of all three species.

Nymphal Habits. Nothing recorded.

Life History. Adults of all species emerge in June (and possibly through August) in Georgia.

Adult Characteristics. Wing length: 5–7 mm. Forewings with marginal intercalaries in pairs. Hind wings extremely minute, with only one partial or no longitudinal veins, and without a costal projection. Male genitalia not illustrated (see Müller-Liebenau, 1974).

Mating Flights. Unknown.

Taxonomy. McDunnough, 1925. Type species: *H. curiosum* (McDunnough); type locality: Ottawa, Ontario. This genus was described by McDunnough, and Edmunds and Traver (1954) synonymized it with *Baetis*. Müller-Liebenau (1974) redescribed the genus as *Rheobaetis* with three species. McCafferty and Provonsha (1975) synonymized *Rheobaetis* with *Heterocloeon*, thus revalidating the genus, and noted that *traverae* Müller-Liebenau is a junior synonym of *curiosum* McDunnough. Adults of *berneri* are still unknown, but Müller-Liebenau (1974) presents a superb taxonomic treatment of the genus with keys to the species of nymphs and adults.

Distribution. Nearctic. Quebec and Ontario south to Georgia, where all three species occur.

Species	United States						Canada			Mexico		Central America
	SE	NE	C	SW	NW	A	E	C	W	N	S	
berneri (Müller-Liebenau)	X	. . .	. . .	. . .	. . .	. . .	. . .	. . .	. . .	. . .	. . .	. . .
curiosum (McDunnough)	X	X	X	. . .	. . .	. . .	X	. . .	. . .	. . .	. . .	. . .
petersi (Müller-Liebenau)	X	. . .	. . .	. . .	. . .	. . .	. . .	. . .	. . .	. . .	. . .	. . .

Paracloeodes Day (figs. 100, 107, 255, 258, 263, 366)

Nymphal Characteristics. Body length: 4 mm. Anterolateral portion of head deeply cut away, exposing mouthparts; outer apical margin of distal segment of labial palpi terminating in a sharp point. Claws slightly curved, about three-fifths as long as respective tarsi, eighteen to twenty minute, slender, straight denticles on inner margin of basal portion (normally difficult to see). Hind wing pad absent. Gills on segments 1–7 single and ovate, with main tracheal trunks pigmented, no lateral branches. Three caudal filaments; terminal filament almost as long as cerci and about equally stout; short hairs on each side of terminal filament and on inner side of cerci. Illustration: Not illustrated in full; parts in Day, 1955, figs. 7–12.

Nymphal Habitat. It is suspected that the nymphs live in the deep waters of large rivers until they reach maturity and migrate into depths of four to six inches near stream margins. They seem to prefer waters with a strong current over fine sand. They are warm-water species and show remarkable ability to survive under marginal conditions. Type locality from which *P. abditus* was described has been dredged, diverted, and dammed and has suffered pollution from irrigation runoff, crop dusting, sewage, and industry. The nymphs have been taken from streams where the water temperatures ranged from 24° C to 27° C (75° F to 81° F).

Nymphal Habits. Not described.

Life History. Length of nymphal life unknown. In laboratory emergence of subimagoes occurred between 8:00 P.M. and 9:00 P.M., and the subimaginal molt occurred the same night. Adults were reared in August.

Adult Characteristics. Wing length: 3.5 mm. Eyes of male turbinate, not contiguous; stalked; turbinate portion large. Metascutellum in lateral view prominent but not flattened; metathoracic postnotum and epimeron deeply emarginate medially (figs. 258, 263). Forewings with marginal intercalaries paired; intercalaries distinctly longer than width of space between members of a pair; hind margin of wing in basal half somewhat evenly rounded. Hind wings absent. Male genitalia illustrated in fig. 366.

Mating Flights. Not described.

Taxonomy. Day, 1955. *Pan-Pac. Entomol*. 31:121. Type species: *P. abditus* Day; type locality: Tuolumne River, Stanislaus County, California. We have examined nymphs of the species described as *Pseudocloeon minutum* Daggy and have transferred the species to *Paracloeodes*.

Distribution. Nearctic and Neotropical (Puerto Rico). The known distribution of this genus is very scattered. In the past the adults were likely to have been regarded as *Pseudocloeon*. It therefore seems probable that other species of the genus have been overlooked in the southern United States, Mexico, or Central or South America.

	United States						Canada			Mexico		Central
Species	SE	NE	C	SW	NW	A	E	C	W	N	S	America
abditus Day	...	...	...	X	...	...	...	...	...	...	...	...
minutus (Daggy)	...	...	X	...	...	...	...	...	...	...	...	...
Species?	X	...	...	...	...	...	...	...	...	...	...	...

Pseudocloeon Klapalek (figs. 111, 254, 257, 261, 367, 401)

Nymphal Characteristics. Body length: 3–6 mm. Body relatively flattened, head somewhat depressed, antennae usually inserted low on head. Distal segment of labial palpi rounded apically. Legs rather long and held out to side. Claws denticulate, the basal denticle always larger; each claw about one-third as long as tarsus. Hind wing pads absent. Gills on abdominal segments 1–7 single and ovate. Two well-developed caudal filaments; terminal filament represented by a minute rudiment. Illustration: Fig. 401.

Nymphal Habitat. Shallow, fairly rapidly flowing water in all sizes of streams and rivers. Nymphs have also been collected along the margins of

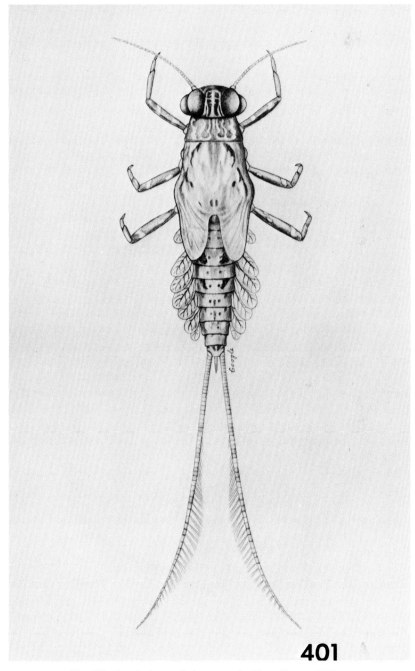

401

Fig. 401. *Pseudocloeon alachua*, nymph. (From Berner, 1950.)

lakes where there is some wave action. The preferred habitat in small streams is the upper side of rocks or other solid structures in a swift current. The nymphs sometimes move into crevices, but mostly they are exposed. In swift water with vegetation the nymphs cling to tips of leaves, fully exposed to the current. Often they can be found in numbers in small pebbly riffles.

Nymphal Habits. Positively rheotropic. The nymphs swim short distances by vigorous abdominal undulations and strong but ineffective lashing of the weakly setaceous caudal filaments. When they stop swimming, they spread the legs, raise the tail, and settle to the bottom. They are well adapted to living in swift waters. The thorax and legs are relatively large, and the abdomen is short and relatively broad. Herbivorous.

Life History. The complete life cycle probably requires six to nine months. Emergence of the subimagoes occurs in the afternoon. The nymph rises to the surface, the skin splits along the dorsum, and the subimago immediately emerges. The subimagoes live from eight to ten hours. In Florida adults appear throughout the year; in other parts of the range adults appear during the spring and summer months on into September.

Adult Characteristics. Wing length: 3.5–6.0 mm. Turbinate eyes of male set on high stalks. Metascutellum in lateral view prominent and flattened, often projecting; metathoracic postnotum and epimeron shallowly emarginate medially (figs. 257, 261). Forewings with stigmatic crossveins slanting and partly anastomosed; marginal intercalaries in pairs; intercalaries distinctly longer than width of space between the members of a pair; hind margin of basal half of forewing rather evenly rounded. Hind wings absent. Abdominal segments 2–6 of male range from whitish to dark olive brown in color; head, thorax, and posterior abdominal terga darker brown. Male genitalia illustrated in fig. 367.

Mating Flights. Males have been observed flying in typical up-and-down pattern over a small North Carolina stream at 5:00 P.M., but no mating was observed. The males rose to a height of about fifteen feet and dropped to within inches of the water. The females did not rise and fall but flew straight forward with occasional dips.

Taxonomy. Klapalek, 1905. *Mitt. Naturh. Mus. Hamburg* 22:105. Type species: *P. kraepelini* Klapalek; type locality: Buitenzorg, Java. The male adults of this genus are generally identifiable to species. There are no workable keys to the nymphs presently available, and it is expected that relatively minute characteristics will have to be used for differentiating the species.

Distribution. Cosmopolitan, except New Zealand. In Nearctic, most common in the east and in western Canada but also occurring south to Florida, Tennessee, Illinois, and Utah. In view of the fact that adults of *Apobaetis*, *Paracloeodes*, and *Baetodes* were all easily confused with *Pseudocloeon*, the

Neotropical species assigned to *Pseudocloeon* are all of uncertain generic affinity. Nymphs assigned by Roback (1966) as genus 1, 2, and 3 near *Pseudocloeon* are allied to *Baetis* and may have adults assignable to *Pseudocloeon*. Thus, besides *Apobaetis*, *Paracloeodes*, and *Baetodes* there are some Neotropical species that might ultimately be placed in *Pseudocloeon*.

Species	United States						Canada			Mexico		Central America
	SE	NE	C	SW	NW	A	E	C	W	N	S	
alachua Berner	X	...	...	...	...	...	...	...	...	...	...	...
anoka Daggy	...	...	X	...	...	...	...	...	...	...	...	...
bimaculatum Berner	X	...	...	...	...	...	...	...	...	...	...	...
carolina Banks	X	X	...	...	...	...	X	...	...	...	...	...
chlorops (McDunnough)	...	...	...	...	...	...	X	...	...	...	...	...
cingulatum McDunnough	...	...	...	...	...	...	X	...	...	...	...	...
dubium (Walsh)	X	X	X	...	...	...	X	...	...	...	...	...
edmundsi Jensen	...	...	...	...	X	...	...	...	...	...	...	...
elliotti Daggy	...	...	X	...	...	...	...	...	...	...	...	...
etowah Traver	X	...	...	...	...	...	...	...	...	...	...	...
futile McDunnough	...	...	...	...	...	...	...	...	X	...	...	...
ida Daggy	...	...	X	...	...	...	...	...	...	...	...	...
myrsum Burks	...	...	X	...	...	...	...	...	...	...	...	...
parvulum McDunnough	X	...	X	...	...	...	X	...	X	...	...	...
punctiventris McDunnough	X	...	X	...	...	...	X	...	...	...	...	...
rubrolaterale McDunnough	...	...	...	...	...	...	...	...	X	...	...	...
turbidum McDunnough	...	...	...	X	...	...	...	...	X	...	...	...
veteris McDunnough	...	...	X	...	...	...	...	...	...	...	...	...
virile McDunnough	...	...	...	...	...	...	X	...	...	...	...	...

Family OLIGONEURIIDAE

This family of Heptagenioidea is almost certainly derived from proto-*isonychiine* Siphlonuridae. The adults of this family are among the most distinctive mayflies known. The nymphs are also readily recognizable, having diverged considerably from their nearest relatives. The Asian genus *Chromarcys*, the most primitive member of the family, is placed in a separate subfamily Chromarcyinae or with fossil genera in the Hexagenitinae, but all other genera are in the subfamily Oligoneuriinae.

Subfamily Oligoneuriinae

Although predominantly pantropical, the Oligoneuriinae extend north in the Old World to Central Europe, Afghanistan, and Japan and in the New World to Saskatchewan. To the south they extend to Natal in Africa, Madagascar, and the Sunda Islands and to Argentina and Peru in South America. Only two genera cross from the Neotropical region into North America.

Nymphal Characteristics. Maxillary and labial palpi two-segmented; a tuft of gills attached at bases of maxillae. Fore femora and tibiae with a double row of long setae on inner side. Gills on abdominal segment 1 ventral, gills on segments 2–7 dorsal.

Adult Characteristics. Forelegs shorter than middle pair. Longitudinal veins of forewings reduced by fusion and loss, only three or four compound longitudinal veins present behind R_1. Two or three caudal filaments; terminal filament vestigial or as long as cerci.

Homoeoneuria Eaton (figs. 114, 196, 368)

Nymphal Characteristics. Body length: 9–12 mm. Body streamlined. Head rounded, eyes lateral. Fore tarsi reduced to small protuberances; mesothoracic and metathoracic claws slender, without denticles. Ventral gills on abdominal segment 1 large and multibranched; gills on segments 2–7 dorsal, small, flat, and slender. Three caudal filaments, approximately equal in length. Illustration: Edmunds, Berner, and Traver, 1958, figs. 15–29.

Nymphal Habitat. Large, moderately rapid to rapid streams with shifting sand bottoms.

Nymphal Habits. The nymphs burrow rapidly into the sand and live about one or two inches below the surface. They are awkward swimmers; when brought above the level of the sand, they often rest on their sides. They filter food from the water in the interstices of the sand with long hairs on their forelegs and graze on the materials so collected.

Life History. Length of nymphal life unknown; probably less than one year from egg to adult. The subimagoes probably molt to the imago stage while on the wing, shedding the cuticle from all parts of the body except the wings. Emergence occurs from June in Florida to October in northern localities.

Adult Characteristics. Wing length: 6–7 mm. Compound eyes large, contiguous, covering most of head in both sexes. Wing venation much reduced, lacking veins R_3 and IR_3, thus only three apparent longitudinal veins behind vein R_1. Hind legs with strongly bowed coxae and femora; short, coarse, curved spines on inner surfaces of femora and tibiae; tarsi terminating in large bulblike claws. Males lack genital forceps. Male genitalia illustrated in fig. 368.

Mating Flights. The flight of *Homoeoneuria* in Florida occurs in midmorning in full sunlight (W. L. Peters, personal communication), but in Indiana the males fly at dusk in the evening (W. P. McCafferty, personal communication). The absence of genital forceps in the male and the shape of the hind legs, with their strongly bowed coxae and femora, indicate that the female must be held by the hind legs during mating. Additional evidence for such a conclusion is the presence of short, coarse, curved spines on the inner surfaces of the femora and the tibiae where they would grip the female and the upturned legs in a position to permit the femora to embrace the female's abdomen. The large, bulblike claws with their pubescent surfaces also may assist the male in holding the female.

Taxonomy. Eaton, 1881. *Entomol. Mon. Mag.* 17:192. Type species: *H. salviniae* Eaton; type locality: Dueñas, Guatemala. This genus has been treated by Edmunds, Berner, and Traver (1958), and the adults of the two species described from the United States are keyed. There are some undescribed species in North America, and the genus unquestionably occurs in Mexico (the species *H. salviniae* Eaton was described from Guatemala); we have also seen specimens from Amazonas, Brazil. The nymphs cannot yet be identified to species.

Distribution. Neotropical and Nearctic. In the Nearctic extending north to South Carolina, Indiana, Nebraska, and Utah.

Species	United States						Canada			Mexico		Central America
	SE	NE	C	SW	NW	A	E	C	W	N	S	
ammophila (Spieth)	...	...	X	...	...	...	...	...	...	...	...	...
dolani Edmunds, Berner, & Traver ...	X	...	...	...	...	...	...	...	...	...	...	...
salviniae Eaton	...	...	...	...	...	...	...	...	...	...	...	X
Species?	...	...	...	X	...	...	...	...	...	...	...	...

Lachlania Hagen (figs. 44, 115, 195, 369, 402)

Nymphal Characteristics. Body length: 8–10 mm. Body somewhat depressed. Head more or less flattened; eyes dorsal; maxillae with tufts of gills at bases. Claws short and stout. Abdomen somewhat flattened with lateral margins prolonged into rather coarse posterolateral projections. Ventral gills on abdominal segment 1 small; gills on segments 2–7 dorsal, plate-like, small. Two caudal filaments with rather sparse fringe of setae on inner margin. Illustration: Fig. 402.

Nymphal Habitat. The nymphs are usually found clinging to small sticks lodged in interstices among rocks in rapids or clinging to the undersides of rocks. In sandy stretches of rivers with considerable current submerged twigs may be literally covered with nymphs.

Nymphal Habits. When disturbed, the nymphs tip the caudal filaments up over the back. They are slow moving and cling to sticks and rocks with great tenacity.

Life History. There is a single generation per year in Utah with length of nymphal life being about three months. Emergence in Utah and New Mexico occurs in August and early September. The adult life is extremely short, perhaps not more than four or five hours.

Adult Characteristics. Wing length: 8–10 mm. Veins R_3 and IR_3 present in forewing, thus the wing has four or more apparent longitudinal veins behind R_1. Males have genital forceps; male genitalia illustrated in fig. 369.

Mating Flights. Mating flights of *L. powelli* take place in midmorning, starting as early as 7:45 A.M., and in *L. dencyanna* from 11:30 A.M. to 1:30 P.M. The flight is very rapid, a characteristic which is probably common to all the Oligoneuriinae but which is unlike that of any other mayflies. The subimaginal skin is shed from all parts of the body except from the wings while the insects are in flight. Mass mating flights occur only on certain days, but the controlling ecological factors are unknown.

Taxonomy. Hagen, 1868. *Proc. Soc. Nat. Hist. Boston*, p. 373. Type species: *L. abnormis* Hagen; type locality: Cuba. Koss and Edmunds (1970) present a key to the three species of *Lachlania* occurring in the United States and Canada, although the nymph of the Canadian species is still unknown. There are probably a number of undescribed species in Mexico and Central America.

Distribution. Neotropical and Nearctic. In the Nearctic the species extend north into Utah, Wyoming, and Saskatchewan.

	United States						Canada			Mexico		Central
Species	SE	NE	C	SW	NW	A	E	C	W	N	S	America
dencyanna Koss	...	...	...	X	...	...	...	...	...	...	...	...
fusca (Navas)	...	...	...	...	...	...	...	...	...	...	...	X
lucida Eaton	...	...	...	...	...	...	...	...	...	...	...	X
powelli Edmunds......	...	...	...	X	X	...	...	...	...	...	...	...
saskatchewanensis Ide	...	...	...	...	...	...	...	X	...	...	...	...
Species?	...	...	...	...	...	...	...	...	...	X	X	...

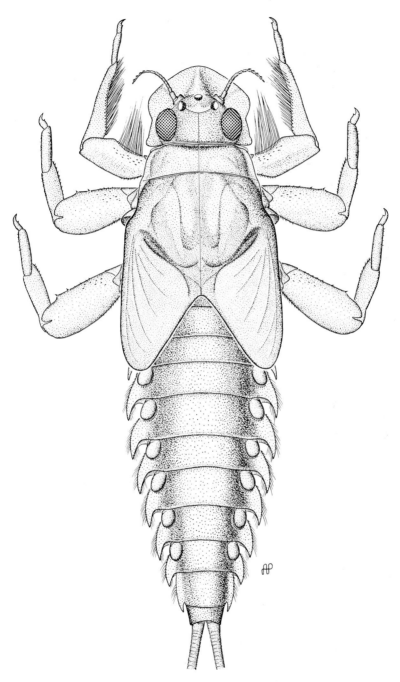

Fig. 402. *Lachlania powelli*, nymph.

Family HEPTAGENIIDAE

The Heptageniidae were almost certainly derived from near the base of the phyletic line which evolved also into *Isonychia*. The nymphs of Arthropleinae have specialized mouthparts, but the adults and nymphs also show many primitive characters; therefore, the genus *Arthroplea* is sometimes placed in a separate family.

The family is diverse and abundant in the Holarctic, especially in the northern regions. A few genera are found in the Ethiopian region, these being closely allied to the more diverse Oriental fauna which extends to the Sunda Islands. The family is absent from the Australian region (the one published record from this area is erroneous). Four Nearctic genera extend to Central America. One wing which is generically unidentifiable is the basis of a record from Brazil.

Nymphal Characteristics. Body distinctly depressed or flattened; head prognathous with eyes and antennae dorsal and the head capsule forming the entire dorsal surface. Maxillary and labial palpi two-segmented (in *Pseudiron*, maxillary palpi three-segmented). Abdominal gills present on segments 1–7 consist of single semioval or rarely slender lamellae usually with fibrilliform tufts at or near the base. Two or three caudal filaments present.

Adult Characteristics. Forewings with two pairs of cubital intercalaries and with veins MP_1 and MP_2 forming a more or less symmetrical fork. Hind legs with tarsi five-segmented (in *Pseudiron*, basal tarsal segment partially fused to tibia). Two caudal filaments present.

Subfamily Heptageniinae

This subfamily includes most of the family, and the geographic range is the same as that for the family.

Nymphal Characteristics. Maxillary palpi less than three times as long as galea-lacinia; setae on palpi not arranged into two distinct rows. Claws much shorter than tibiae. Abdominal gills inserted dorsally or laterally but may extend beneath the abdominal venter.

Adult Characteristics. Vein MA of hind wings usually forked (simple in *Cinygma dimicki*, in which veins R_{2+3} and IR_3 are detached basally). Fore tarsi longer than fore tibiae; hind tarsi with five clearly differentiated segments. Genital forceps with two short terminal segments.

Cinygma Eaton (figs. 131, 271, 388)

Nymphal Characteristics. Body length: 10–12 mm. Head one and one-fourth times as wide as long; width of labrum less than one-eighth width of head; apex of galea-lacinia of maxillae with long setae. Gills not extending beneath abdominal venter; lamella of gill 1 about half as long as those of gills 2–7, fibrilliform portion of gill 1 twice as long as lamella. Three caudal filaments. Illustration: Not illustrated in full; parts in McDunnough, 1933, pl. 2.

Nymphal Habitat. In Sierra Nevada nymphs occur at altitudes of 3,000 to 7,000 feet in moderately fast water, nearly always clinging to wood and bark of dark color. In Idaho they occur usually at altitudes above 5,000 feet in moderately fast streams, most commonly under large flat rocks and on submerged logs. The nymphs are found at low elevations in the coastal streams of the Pacific Northwest.

Nymphal Habits. The nymphs move very sluggishly, and they are poor swimmers.

Life History. Unknown. Adults have been recorded from early May to late July.

Adult Characteristics. Wing length: 11–12 mm. Eyes of male large, ovoid, separated dorsally by space twice width of median ocellus. Forelegs of male as long as body; tarsi twice length of tibiae; basal tarsal segment two-thirds to seven-eighths as long as second segment. Stigmatic crossveins of forewing numerous and anastomosed so that the fine vein parallel with the costal vein divides costal interspace into two series of cellules; basal costal crossveins weak. Subanal plate of female broadly rounded apically with median notch. Male genitalia as in fig. 388.

Mating Flights. In Idaho *Cinygma dimicki* has been observed swarming in late afternoon and early evening over smooth water. The males were only one to three feet above the stream and flew with short and slow vertical motions, usually hovering in one place. The males were spaced at least three feet apart.

Taxonomy. Eaton, 1885. *Trans. Linn. Soc. London*, 2d Ser. Zool., 3:247. Type species: *C. integrum* Eaton; type locality: Mount Hood, Oregon. The males of this genus are readily identifiable to species on the basis of color pattern and genitalia. The nymphs cannot be identified to species.

Distribution. Holarctic and Oriental. In the Nearctic, the species occur in northwestern North America south to California, Nevada, Idaho, and Wyoming.

Species	United States						Canada			Mexico		Central America
	SE	NE	C	SW	NW	A	E	C	W	N	S	
dimicki												
McDunnough.......	...	...	...	...	X	...	...	...	...	...	...	...
integrum Eaton	...	...	...	...	X	...	...	...	X	...	...	...
lyriformis												
(McDunnough)	...	...	...	...	X	...	...	...	X	...	...	...
Species?.............	...	...	...	X	...	...	...	...	...	...	...	...

Cinygmula McDunnough (figs. 128, 129, 268, 270, 276, 287)

Nymphal Characteristics. Body length: 7–11 mm. Head 1 1/5 times as wide as long; frontal margin of head capsule with distinct median emargination; width of labrum about one-fourth width of head; apex of galea-lacinia of maxillae with comblike spines; maxillary palpi normally partially visible at sides of head in dorsal view. Gills similar on all segments; fibrilliform portion reduced to two or three filaments or absent; gills on abdominal segment 1 extend only slightly beneath abdomen or not at all. Three caudal filaments. Illustration: Not illustrated in full; parts in McDunnough, 1933, pl. 3.

Nymphal Habitat. One of the most common mayflies of the western mountains. The nymphs live under stones in all parts of many streams. In Colorado and Utah they have been found in some of the high streams and lakes (8,000 to 11,000 feet) which have clean rocky bottoms. In streams of California nymphs have been found in water only one or two inches deep at the foot of waterfalls; they live in crevices and on the lower surfaces of small stones in streams of the redwood belt of the Coast Ranges and the Sierra Nevada.

Nymphal Habits. The nymphs can move fairly rapidly across stones when disturbed but are very poor swimmers. They are commonly found drifting in streams.

Life History. In one coastal stream in Oregon *Cinygmula reticulata* eggs hatch in fall, and the nymphs grow slowly in winter, increasing the growth rate in early spring. The adults emerge from April to June. The nymphs take nine to ten months to develop. The same species in the Metolius River, a large spring-fed stream, however, has all size classes present nearly all the time, and the adults emerge from April to November. The life cycle in this river is apparently uncorrelated because of the lack of seasonal temperature signals necessary to coordinate the life cycle. The adults of other species emerge from April to October, but in almost all parts of the range the adults of any one species emerge over a period of sixty to seventy-five days or less.

Adult Characteristics. Wing length: 7–10 mm. Eyes of male large, separated dorsally by space ½ to 1½ times width of median ocellus. Wing membrane often tinged with gray, yellow, or amber; crossveins of stigmatic area usually not anastomosed. Forelegs of male longer than body; forelegs of female about as long as body; fore tarsi of male 1¼ to 1½ times as long as tibia; basal segment of fore tarsi ⅔ to 5/6 as long as second segment. Subanal plate of female with deep V-shaped median emargination. Male genitalia illustrated in fig. 270.

Mating Flights. The swarms consist of numerous adults above streams. Some species fly in sunlight at midday, but most species swarm in the evening. Mating flights often occur earlier on cloudy days, in the shade, or late in the season. Most species fly five to fifteen feet above the water with individuals rising and falling about two feet. The males appear to live long enough to swarm more than once. Egg laying was observed once in the early morning above a riffle; the eggs were discharged in a cluster before the female touched the water surface. On reaching the water the eggs were scattered. After resting briefly, the female extruded a second mass of eggs. The female repeated the performance of egg laying four times before being spent.

Taxonomy. McDunnough, 1933. *Can. Entomol.* 65:75. Type species: *C. ramaleyi* (Dodds); type locality: Tolland, Colorado. The adults of this genus can generally be identified to species on the basis of the male genitalia. Wing coloration is highly variable within a species. When the wing is well colored, the pattern provides a valuable aid in the identification of the species, but the wing color must be assessed with care because the characteristic is subject to modification depending on the temperature of the water in which the nymph develops. The nymphs usually cannot be identified to species. Many species have been reared, but no readily usable characters have been found to distinguish them. A single species has been described from eastern North America, and all nymphs seem to be of one type with the terga generally much paler than in the species from western North America.

Distribution. Holarctic. In the Nearctic most species occur in western North America south to California and New Mexico, with one eastern species south to North Carolina.

Species	United States						Canada			Mexico		Central America
	SE	NE	C	SW	NW	A	E	C	W	N	S	
confusa (McDunnough)	...	...	...	...	...	...	...	...	X	...	...	...
gartrelli McDunnough.......	...	...	...	...	...	...	...	...	X	...	...	...
kootenai McDunnough.......	...	...	...	X	...	...	...	...	X	...	...	...
mimus (Eaton)	...	...	...	X	X	...	...	...	X	...	...	...
par (Eaton)	...	...	...	X	X	...	...	...	X	...	...	...
ramaleyi (Dodds)	...	...	...	X	X	...	...	...	X	...	...	...
reticulata McDunnough.......	...	...	...	...	X	...	...	...	...	...	...	...
subaequalis (Banks)	X	X	...	...	...	...	X	...	...	...	...	...
tarda (McDunnough)	...	...	...	X	X	...	...	...	X	...	...	...
tioga Mayo	...	...	...	X	...	...	...	...	...	...	...	...
uniformis McDunnough.......	...	...	...	X	...	...	...	...	X	...	...	...
Species?..............	...	...	...	...	...	X	...	...	...	...	...	...

Epeorus Eaton

Nymphal Characteristics. Body length: 7–18 mm. Frontal and anterolateral margins of head expanded, with dense marginal setae; head 1⅓ to 1½ times as wide as long; labrum ¼ to 1/5 as wide as head; inner apical margin of galea-lacinia with triad of stout spines, each slightly incurved and slender at apex. Gills composed of platelike lamellae and a reduced fibrilliform portion; in many species anterior lobe of lamellae of gill 1 well developed; in some species anterior lobes meeting or almost meeting beneath the body; gill 7 may or may not project beneath abdominal venter to form, along with intermediate gills, a partial or complete disc. Without submedian spines on abdominal terga. Two caudal filaments. Illustration: fig. 403.

Nymphal Habitat. The nymphs live in shallow, cool or cold, rapidly flowing water, where they attach themselves to rocks, sticks, or other firmly anchored material. In the Rocky Mountains they are found in small to medium-sized streams at elevations of 4,000 to 10,000 feet, but in the streams that border the Pacific coast they may be found down to sea level.

Nymphal Habits. The nymphs, especially those whose gills form a disc, hold on to rocks tenaciously when attempts are made to remove them. They are poor swimmers.

Life History. The length of time required for development in three species is seven or eight months. The adults have been observed emerging in early morning and late afternoon. The nymphs rise to the surface, and the subimagoes immediately escape from the turbulent water. The nymphs have been observed crawling to within an inch of the surface of the water, emerging there from the nymphal skin, and breaking through the surface film with wings fully expanded. The adults of various species emerge from April to September.

Adult Characteristics. Wing length: 7–19 mm. Eyes of male large, usually contiguous dorsally. Forelegs of male as long as or slightly longer than body; tibiae about three-fourths length of tarsi; basal fore tarsal segment slightly shorter than, subequal to, or longer than second segment. Subanal plate of female with moderate to shallow emargination. Male genitalia illustrated in figs. 269, 279.

Mating Flights. There is considerable variation in the pattern observed in the Rocky Mountain species. *E. albertae* flights consist of single individuals or small swarms of males that fly over stream riffles in direct sunlight in the morning and early evening. This species has a steady flight with little up-and-down motion. *E. deceptivus* adults have been reported swarming in bright sunlight at about 4:00 P.M. in early September in Utah. *E. longimanus* forms medium-sized to large swarms during the evening or on cloudy days. The swarming of this species seems to be directly correlated with the amount of

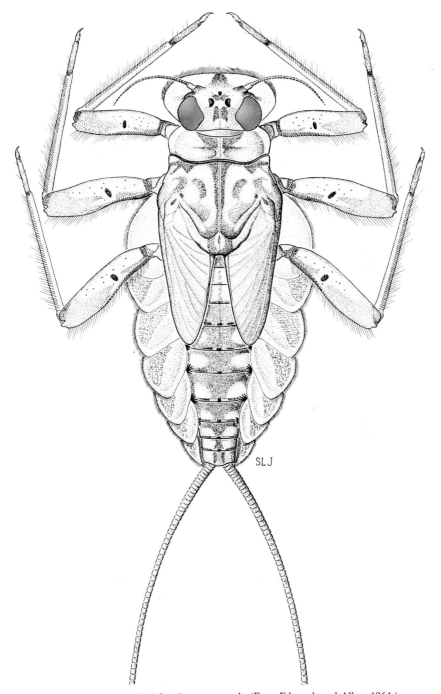

Fig. 403. *Epeorus* (*Iron*) *longimanus*, nymph. (From Edmunds and Allen, 1964.)

light; the reduction of light to below 600 footcandles seems to trigger the activity. At about 80 footcandles swarming seems to reach its peak, followed by a rapid decline in the number of swarming individuals as the light intensity drops. At 20 footcandles the swarm is completely dispersed. It appears that *E. longimanus* males are comparatively long-lived and that they may participate in swarms on at least two days.

Taxonomy. Eaton, 1881. *Entomol. Mon. Mag.*, 18:26. Type species: *E. torrentium* Eaton; type locality: Tarascon, France. The adults should be quite readily keyed to species except with perhaps some confusion in the northeastern species. Some of the nymphs may be identified to species provided caution is used in applying names. Some local or regional keys to species are available.

Distribution. Holarctic, Oriental, and Neotropical. Widespread in the Nearctic south into Central America to Honduras.

Subgenus *Iron* Eaton (figs. 267, 275, 279, 403)

Nymphal Characteristics. Body length: 7–14 mm. Frontal and lateral margins of head moderately expanded; head 1⅓ to 1½ times as wide as long. Anterior lobe of gill 1 and posterior portion of gill 7 variable; may partially or completely extend beneath the abdomen to form a ventral disc. Body and gills usually pale to brown in color. Abdominal terga with scattered setae or with moderate medial row of setae. Illustrations: Fig. 403.

Adult Characteristics. Wing length: 7–11 mm. Basal fore tarsal segment longer than second segment. Stigmatic crossveins of forewing slanted; basal costal crossveins weak, appearing detached anteriorly. Male genitalia as in fig. 279.

Taxonomy. Eaton, 1883. *Trans. Linn. Soc. London*, 2d Ser. Zool., 3, pl. 24. Type species: *E. (I.) longimanus* (Eaton); type locality: Manitou, Colorado. The Rocky Mountain species of this subgenus can be easily identified using the keys of Edmunds and Allen (1964). Species (and especially the nymphs) in the remainder of North America are not as easily identified.

Distribution. Holarctic, Neotropical, and Oriental. Widespread in the Nearctic south into Central America to Honduras.

Species	United States						Canada			Mexico		Central America
	SE	NE	C	SW	NW	A	E	C	W	N	S	
albertae (McDunnough)	...	...	...	X	X	...	...	...	X	...	...	...
deceptivus (McDunnough)	...	...	...	X	X	...	...	...	X	...	...	...
dispar (Traver)	X	...	...	...	...	...	...	...	...	...	...	...
dulciana (McDunnough)	...	...	...	X	...	...	...	...	X	...	...	...

Species	United States						Canada			Mexico		Central America
	SE	NE	C	SW	NW	A	E	C	W	N	S	
fragilis (Morgan)		X					X					
frisoni (Burks)		X										
hesperus (Banks)				X								
lagunitas (Traver)				X								
longimanus (Eaton)				X	X	X			X			
margarita Edmunds & Allen				X								
metlacensis Traver											X	
namatus (Burks)			X									
pleuralis (Banks)	X	X					X					
punctatus (McDunnough)		X					X					
rubidus (Traver)	X											
subpallidus (Traver)	X											
suffusus (McDunnough)							X					
vitreus (Walker)		X	X				X					
Species?												X

Subgenus *Ironopsis* Traver (figs. 117, 265, 269, 274, 288)

Nymphal Characteristics. Body length: 12–18 mm. Frontal and anterolateral margins of head greatly expanded; head 1½ times as wide as long. Anterior lobe of gill 1 and posterior portion of gill 7 extend beneath the venter of body to form a ventral disc. Gill lamellae usually purplish brown in color. Prominent ridge of long setae present along the middorsal line of all abdominal terga. Illustrations: Kapur and Kripalani, 1963, fig. 14.

Adult Characteristics. Wing length: 13–19 mm. Basal fore tarsal segment slightly shorter or subequal to second segment. Stigmatic crossveins of forewing usually anastomosed; basal costal crossveins well developed. Male genitalia as in fig. 269.

Taxonomy. Traver, 1935. *Can. Entomol.* 67:36. Type species: *E. (I.) grandis* (McDunnough); type locality: Waterton Lakes, Alberta. The adults of this subgenus should be readily identifiable to species, but no published keys exist to differentiate the nymphs of the two North American species. *E. (I.) grandis* is the more common and widespread species in the subgenus.

Distribution. Holarctic. The two Nearctic species are restricted to northwestern North America south to California, Nevada, and Wyoming. An apparently undescribed species occurs in Vera Cruz (Mexico), according to R. K. Allen (personal communication). *E. (I.) permagnus* is known from Oregon.

Heptagenia Walsh (figs. 53, 120, 121, 132, 235, 284, 286, 296)

Nymphal Characteristics. Body length: 6–12 mm. Head 1⅓ to 1½ times as wide as long; labrum ⅜ to ½ width of head; apex of galea-lacinia of maxillae with comblike spines. Gills on abdominal segments 1–7 usually similar but not same size; in some species fibrilliform tuft greatly reduced or lacking on seventh pair; gill 7 in some species with lamellae narrow (minute in one species); gills not extending beneath abdominal venter. Three caudal filaments. Illustration: Burks, 1953, fig. 383.

Nymphal Habitats. Generally the nymphs occur under stones and among debris in shallow, moderately rapid to rapid water near the banks of streams and rivers. The mature nymphs often occur under stones at the edge of quiet pools. They also occur at the margins of lakes in which there is wave action along rocky shores and in debris to which the insects may attach themselves.

Nymphal Habits. Very active; when a stone is lifted from the water, the nymphs often scurry to the underside. They are agile and able to move equally well to front, back, or side. They cling closely and with considerable strength to the surface to which they are attached.

Life History. Hatching in the laboratory takes from twelve to forty days. The length of nymphal life is probably one year or less, although this has not been determined precisely; the nymphal life for *H. elegantula* in Utah is apparently less than seventeen weeks. When ready to emerge, the nymph rises to the surface and the subimago bursts free. Some subimagoes emerge from nymphs attached to rocks beneath the water. The adult life lasts from two to four days. Adults have been taken from April to October. The principal emergence occurs in June and July.

Adult Characteristics. Wing length: 5–15 mm. Wide variation in body color from creamy white to dark purplish or reddish brown. Eyes of male large, contiguous dorsally or separated by less than width of median ocellus. Forelegs of male slightly longer than body; tarsi about 1¼ times length of tibiae; basal fore tarsal segment varies from 1/6 to ½ length of second segment. Basal costal crossveins of forewing usually well developed; stigmatic crossveins variable but usually not anastomosed; in some species costal and subcostal crossveins thickened or margined with brown; sometimes crossveins in several spaces below bulla also margined and crowded together. Subanal plate of female rounded apically or with small median indentation. Male genitalia illustrated in fig. 284.

Mating Flights. The swarming of *H. criddlei* occurs in midmorning or late afternoon in full sunlight from two to six feet above the water or stream bank;

H. elegantula swarms in early evening at a height of ten to twenty feet. *Heptagenia* females scatter their eggs, a few at a time. After mating the females fly singly or in small swarms up and down over streams or the edges of lakes from two to ten feet above the water. In *H. hebe* the flight resembles the mating flight of the males. From time to time a female will drop to the water, touch the surface long enough to release some of her eggs, and then rise again to fly with the other females. In some species the female flies singly in a straight line, occasionally touching down to drop her eggs; then she rises a short distance, reverses directions, and touches the water again, repeating the performance until all the eggs are released. The spent females fly to nearby bushes or trees to rest. In *H. hebe* contact with water seems to be necessary to start the flow of eggs, and the flow continues only so long as the female is in contact with the water.

Taxonomy. Walsh, 1863. *Proc. Entomol. Soc. Philadelphia* 2:197. Type species: *H. flavescens* (Walsh); type locality: Rock Island, Illinois. *Heptagenia* is a diverse genus consisting of several complexes of species. Most species should be recognizable as adults, but some may be difficult to distinguish. The nymphs of many of the species have not yet been described, but several clear-cut complexes can be identified from the existing keys; the regional keys to the species of both adults and nymphs will make many identifications possible. *H. mexicana* Ulmer may be assignable to *Stenonema*. Flowers and Hilsenhoff (1975) have keyed nymphs and adults of the Wisconsin species.

Distribution. Holarctic, Neotropical, and Oriental. Widespread in the Nearctic, extending into Central America as far south as Guatemala.

	United States						Canada			Mexico		Central
Species	SE	NE	C	SW	NW	A	E	C	W	N	S	America
adequata McDunnough	...	...	...	...	...	...	...	...	X	...	...	...
aphrodite McDunnough	X	X	...	...	...	...	X	...	...	...	...	...
criddlei McDunnough	...	...	...	X	X	...	...	...	X	...	...	...
cruentata Walsh	...	...	X	...	...	...	...	X	...	...	...	...
diabasia Burks	...	...	X	...	...	...	...	X	...	...	...	...
dolosa Traver	X	...	...	...	...	...	...	...	...	...	...	...
elegantula (Eaton)	...	...	...	X	X	...	...	X	...	...	...	...
flavescens (Walsh)	X	...	X	...	...	...	...	X	...	...	...	...
hebe McDunnough	...	X	X	...	...	...	X	...	...	...	...	...
horrida McDunnough	...	...	...	...	...	...	X	...	...	...	...	...

Species	United States						Canada			Mexico		Central America
	SE	NE	C	SW	NW	A	E	C	W	N	S	
inconspicua McDunnough			X					X				
jewetti Allen					X							
julia Traver	X											
juno McDunnough	X	X					X					
kennedyi McDunnough				X	X							
lucidipennis (Clemens)		X	X				X					
maculipennis Walsh	X		X					X				
manifesta (Eaton)			X									
marginalis Banks	X	X	X									
mexicana Ulmer											X	
minerva McDunnough	X	X					X					
otiosa McDunnough					X							
patoka Burks			X									
perfida McDunnough							X					
persimplex McDunnough			X									
petersi Allen				X	X							
pulla (Clemens)		X	X				X					
rodocki Traver					X							
rosea Traver				X	X							
rusticalis McDunnough		X	X				X					
salvini Kimmins										X		
simplicioides McDunnough				X	X				X			
solitaria McDunnough				X	X				X			
spinosa Traver	X											
thetis Traver	X											
townesi Traver	X											
umbricata McDunnough			X				X					
walshi McDunnough							X					
Species?												X

Ironodes Traver (figs. 116, 266, 273, 389)

Nymphal Characteristics. Body length: 7–14 mm. Anterolateral and frontal margins of head slightly expanded; frontal margin slightly emarginate at median line, densely fringed with setae; head 1 1/6 times as wide as long; labrum ¼ as wide as head. Galea-lacinia of maxillae lacks triad of stout spines at apex; slender straight spines present. Gills not extending beneath abdominal venter; fibrilliform portion of gills moderately well developed, lamellate portions rather small, obovate. Paired submedian spines on posterior margins of abdominal terga 1–9. Two caudal filaments. Illustration: Traver, 1935, fig. 110.

Nymphal Habitat. Swiftest mountain streams; the nymphs cling to the underside of stones, sticks, or leaf debris where the water can pass over them continuously.

Nymphal Habits. Poor swimmers. The nymphs are strongly thigmotactic. They feed chiefly on detritus.

Life History. It has been suggested that there is more than one brood per year in California because mature nymphs were collected in September and adults were reared in April. In Oregon *I. nitidus* takes about eight months to develop, and the adults emerge from April to July.

Adult Characteristics. Wing length: 8–14 mm. Eyes of male small to medium-sized, separated dorsally by space twice width of median ocellus. Forelegs of male slightly longer than body; fore tarsi 1½ times length of tibiae; basal segment of fore tarsi slightly longer than or subequal to second segment. Basal costal crossveins strong; stigmatic crossveins simple, numerous, somewhat slanting. Female subanal plate with moderately deep median emargination. Male genitalia illustrated in fig. 389.

Mating Flights. Not described.

Taxonomy. Traver, 1935. *Can. Entomol.* 67:32 (misprint as *Ironopsis*). Type species: *I. nitidus* (Eaton); type locality: Mt. Hood, Oregon. The adults can be identified to species with certainty only with much caution, and the nymphs cannot be specifically determined.

Distribution. Nearctic. Restricted to northwestern North America south to California, Nevada, Idaho, and Montana.

| | United States | | | | | | Canada | | | Mexico | | Central |
Species	SE	NE	C	SW	NW	A	E	C	W	N	S	America
arctus Traver.........	...	...	...	...	X	...	...	...	X	...	...	...
californicus (Banks)	...	...	...	X	...	...	...	...	...	...	...	...
flavipennis Traver	...	...	...	...	...	...	...	...	X	...	...	...

Species	United States						Canada			Mexico		Central America
	SE	NE	C	SW	NW	A	E	C	W	N	S	
geminatus (Eaton)	...	...	...	X	X	...	...	...	...	...	...	...
lepidus Traver	...	...	...	X	...	...	...	...	...	...	...	...
nitidus (Eaton)	...	...	...	X	X	...	...	...	...	...	...	...

Rhithrogena Eaton (figs. 118, 272, 285, 371)

Nymphal Characteristics. Body length: 5–12 mm. Frontal margin of head with median emargination; head 1¼ to 1⅓ times as wide as long; width of labrum about ¼ width of head; apex of galea-lacinia of maxillae with comb-like spines. Gills on abdominal segments 1–7 composed of platelike lamellae and tuft of filaments; lamellate portions of gills 1 and 7 meet beneath body; margins of gills 2–6 overlap one another, free edge of each being deflected and pressed against surface to which nymph clings, forming disc on venter. Three caudal filaments. Illustration: Burks, 1953, fig. 390; Eaton, 1883–88, pl. 54.

Nymphal Habitat. The nymphs attach themselves to stones in swift to moderate currents; they are often found on irregularly shaped smooth stones and in water several inches deep. In western mountain localities the nymphs occur at elevations up to 10,400 feet.

Nymphal Habits. When a rock is lifted from the water, the nymphs may move quickly from the upper side to the lower side or may slip into a crevice. Others may remain perfectly still until an attempt is made to remove them. They adhere so tightly to the rocks that they can be removed only with difficulty.

Life History. Development from egg to adult requires about seven months. In Oregon nymphs of *R. morrisoni* hatch in August or September and begin to emerge the next spring about March 1. The nymphs grow slowly from November to February. The nymphs rise to the surface of the water to emerge. The subimagoes of most species emerge in the evening and require about two days to transform into adults. The adults of various species occur from late February to September.

Adult Characteristics. Wing length: 6–15 mm. Adults usually brownish or reddish brown. Eyes of male large, contiguous dorsally. Forelegs of male somewhat longer than body; tarsi 1¼ to 1½ times length of tibiae; basal tarsal segment 1/5 to ⅓ length of second segment. Crossveins of stigmatic area of forewings strongly to weakly anastomosed (not true of some Palearctic species); basal costal and subcostal crossveins weak. Female subanal plate broadly rounded. Male genitalia illustrated in fig. 371.

Mating Flights. The males of *R. impersonata* swarm in great numbers in late May and early June in Michigan. The females enter the swarm and pair immediately. Mating flights form shortly before sunset at a height of about twenty feet and rise with increasing darkness. The western species usually fly from late afternoon to dusk, but they sometimes swarm in midmorning. Males of *R. morrisoni* have been seen swarming at heights of eight to fifteen feet and sometimes at heights of forty to sixty feet. One male of *R. morrisoni* was collected *in copula* with a female of *Ephemerella tibialis*.

Taxonomy. Eaton, 1881. *Entomol. Mon. Mag.* 18:23. Type species: *R. semicolorata* (Curtis); type locality: Europe. The adult males can be distinguished on the basis of differences in the genitalia. Separation of the nymphs to species is difficult, even though some species are very distinctive. Flowers and Hilsenhoff (1975) have keyed nymphs and adults of the Wisconsin species.

Distribution. Holarctic, Neotropical, and Oriental. Widespread in the Nearctic south to Guatemala.

Species	United States						Canada			Mexico		Central America
	SE	NE	C	SW	NW	A	E	C	W	N	S	
amica Traver	X	X	...	...	...	...	...	...	...	...	...	...
anomala McDunnough	...	X	...	...	...	...	X	...	...	...	...	...
brunneotincta McDunnough	...	...	...	...	...	...	X	...	...	...	...	...
decora Day	...	...	...	X	...	...	...	...	...	...	...	...
exilis Traver	X	...	...	...	...	...	...	...	...	...	...	...
fasciata Traver	X	...	...	...	...	...	...	...	...	...	...	...
flavianula (McDunnough)	...	...	...	X	...	...	...	...	...	...	...	...
fuscifrons Traver	X	...	...	...	...	...	...	...	...	...	...	...
futilis McDunnough	...	...	...	X	X	...	...	...	X	...	...	...
gaspeensis McDunnough	...	...	...	...	...	...	X	...	...	...	...	...
hageni Eaton	...	...	...	X	X	...	...	...	X	...	...	...
impersonata (McDunnough)	...	...	X	...	...	...	X	...	...	...	...	...
jejuna Eaton	...	...	X	...	...	...	...	X	...	...	...	...
morrisoni (Banks)	...	...	...	X	X	...	...	...	X	...	...	...
pellucida Daggy	...	...	X	...	...	...	...	...	...	...	...	...
robusta Dodds	...	...	...	X	X	...	...	...	X	...	...	...
rubicunda Traver	X	...	...	...	...	...	...	...	...	...	...	...

Species	United States						Canada			Mexico		Central America
	SE	NE	C	SW	NW	A	E	C	W	N	S	
sanguinea												
Ide*	...	...	X	...	...	...	X	...	...	...	...	...
uhari Traver	X	...	...	...	...	...	...	...	...	...	...	...
undulata												
Banks	...	...	X	X	X	...	...	...	...	...	...	...
virilis												
McDunnough	...	...	...	...	...	...	...	...	X	...	...	...
Species?	...	...	...	...	...	X	...	...	...	...	...	X

*Flowers and Hilsenhoff (1975) have indicated that *Rhithrogena sanguinea* is a junior synonym of *R. impersonata*.

Stenacron Jensen (figs. 119, 122, 123, 277, 280, 281)

Nymphal Characteristics. Body length: 8–14 mm. Head 1⅓ to 1½ times as wide as long; width of labrum almost ½ to almost ⅔ width of head; galea-lacinia of maxillae with comblike armature at apex. Gills not extending beneath abdominal venter; gills 1–6 with apex pointed; gill 7 reduced to slender filament. Abdomen with moderately developed, acute posterolateral projections on segments 8–9. Three caudal filaments. Illustration: Eaton, 1883–88, pl. 57.

Nymphal Habitat. The nymphs are found in moderately rapid to swift rivers and streams where they occur beneath rocks, amid debris, or in vegetation. They also occur on clean shores of lakes in which there is some wave action and objects to which the nymphs can cling.

Nymphal Habits. Similar to *Heptagenia*. The nymphs are awkward swimmers and move with an undulating motion of the body.

Life History. Of the species investigated, development takes place in one year. Under some conditions, however, two broods may occur. In *S. canadense* embryonic development proceeds without diapause, although the time of development is affected by the water temperature. Ide (1935) determined that the eggs of three females of that species, all deposited at the same time and maintained under identical conditions, continued to hatch over a period of six weeks. Nymphal growth is continuous, but it is slow in the winter months. Although the exact number of nymphal instars is not known, Ide has determined that *S. canadense* undergoes at least forty and probably forty-five instars. The nymphs are the overwintering stage. Emergence occurs while the nymph is floating on the surface. The period of emergence is apparently quite long for each species, and in Florida adults of *S. interpunctatum* emerge every month.

Adult Characteristics. Wing length: 8–12 mm. Eyes of male moderately large, usually separated dorsally by width greater than median ocellus; in some species widely separated, up to five or six times width of median ocellus. Length of forelegs of male variable, shorter to much longer than body; fore tarsi longer than tibiae; basal tarsal segment ¼ to ⅔ (in most species ½ to ⅔) length of second segment. Basal costal crossveins of forewing well developed; crossveins between veins R_1 and R_2 below bullae thickened with black pigmentation, two or three crossveins connected or nearly connected with pigmentation. Subanal plate of female rounded at apex, sometimes with small median indentation. Penes of male usually with well-developed lateral cluster of spines, small in several species. Male genitalia illustrated in figs. 280, 281.

Mating Flights. The swarms usually form in the evening with the males rising and falling with little horizontal movement. The height of the swarm varies with the species. Some species swarm six to eight feet above the water. Leonard and Leonard (1962) reported *canadense* as swarming at twenty to fifty feet. Oviposition occurs as the female flies low over the water, touching her abdomen to the water periodically. The eggs are released a few at a time.

Taxonomy. Jensen (1974) *Proc. Entomol. Soc. Wash.* 76:225. Type species: *S. interpunctatum* (Say); type locality: St. Martin's Falls, Albany River, Ontario. The species of this genus present perplexing taxonomic problems. Lewis (1974) has revised the genus (as the *interpunctatum* group of *Stenonema*). He does not attempt to separate subspecies of *S. interpunctatum*. The pattern of variation in this species does not fit the normal subspecies concept, and until the biological nature of the variation is understood attempts to apply names below the species level seem premature. The revision by Lewis (1974) should allow nymphs and adults of the species to be determined. Because of the past confusion of species identification in this genus we have recorded only the distribution of each species and subspecies indicated by Lewis (1974).

Distribution. Nearctic. The genus is restricted to eastern and central North America south to Florida, west to Arkansas and Minnesota.

Species	United States						Canada			Mexico		Central America
	SE	NE	C	SW	NW	A	E	C	W	N	S	
candidum (Traver)	...	...	X	...	...	...	...	X	...	...	...	...
carolina (Banks)	X	X	...	...	...	...	...	...	...	...	...	...
floridense (Lewis)	X	...	...	...	...	...	...	...	...	...	...	...
gildersleevei (Traver)	X	...	X	...	...	...	...	...	...	...	...	...
interpunctatum canadense (Walker)	...	X	X	...	...	...	X	X	...	...	...	...
interpunctatum frontale (Banks)	...	X	...	...	...	...	X	...	...	...	...	...

Species	United States						Canada			Mexico		Central America
	SE	NE	C	SW	NW	A	E	C	W	N	S	
interpunctatum heterotarsale (McDunnough)			X				X					...
interpunctatum interpunctatum (Say)	X		X									...
minnetonka (Daggy) ...			X									...
pallidum (Traver)	X											...

Stenonema Traver (figs. 124–127, 278, 282, 283, 404)

Nymphal Characteristics. Body length: 6–20 mm. Head 1¼ to 1½ times as wide as long; width of labrum almost ½ to almost ⅔ width of head; galea-lacinia of maxillae with setae and/or plumose spines. Gills not extending beneath abdominal venter; gills 1–6 with apex rounded or truncate, gill 7 reduced to slender filament. Abdomen with small to well-developed posterolateral projections on segments 5–9, 6–9, or 7–9. Three caudal filaments. Illustration: Berner, 1950, pl. II.

Nymphal Habitat. The nymphs are found in a variety of habitats, generally in moderately rapid to swift streams of all sizes; they cling to the under surfaces of stones and rocks (where they tend to hide in crevices) or debris lodged in the stream, or they hide among the leaves that accumulate in quiet portions of streams and in pools and backwaters. They may also be found on rooted vegetation at the banks of streams. Some species occur in lakes where there is some wave action near shore; the nymphs attach themselves to debris or stones along the edge.

Nymphal Habits. The nymphs cling tenaciously to the objects on which they live within the streams. They swim awkwardly by undulating the body. Although they swim only in a forward direction, they can walk in almost any direction with equal ease. They are strongly thigmotactic; if several nymphs are placed together in a dish of water with nothing to which they can attach themselves, they gather in a clump and remain thus until an object is placed in the water with them. In its natural environment the nymph crawls into a crevice until the dorsum of the body is in contact with some object. The strongly flattened femora and the spreading legs of the nymph reduce its resistance to the stream flow to a minimum.

Life History. The time required for development has not been determined. In the southeastern United States development may be completed in less than one year; in colder areas of North America it probably requires one year. In several species the development of the eggs takes from eleven to twenty-three

Fig. 404. *Stenonema smithae*, male. (From Berner, 1950.)

days in the laboratory. Thigmotactic and phototactic responses are present in the nymphs upon hatching. Emergence occurs while the nymph floats at the surface of the water and takes only a few seconds; the subimago almost immediately flies away. The subimaginal stage lasts from twenty to thirty-six hours. The period of emergence is fairly long; even in cold northern streams adults have been taken in early April (Illinois) and in early October (Michigan).

Adult Characteristics. Wing length: 6–16 mm. Eyes of male moderately large; usually separated dorsally by space ⅓ width of median ocellus to almost 3 times width of median ocellus. Forelegs of male slightly shorter to slightly longer than body; fore tarsi longer than tibiae; basal tarsal segment ¼ to 5/6 (in most species ½ to ⅔) length of second segment. Claws dissimilar on all tarsi. Basal costal crossveins of forewing well developed; in some species crossveins near bullae crowded. Male genitalia illustrated in fig. 283.

Mating Flights. Just before dusk small bands of ten to twenty males swarm directly over a stream; the height varies with the species, ranging from six to fifteen feet above the water. *S. vicarium* has been reported by Leonard and Leonard (1962) as swarming twenty to fifty feet above streams in Michigan. The males rise and fall as little as two feet or as much as six feet with little horizontal movement. Swarming seems to be directly correlated with light intensity; once swarming has started, the males can be observed in flight until there is no longer sufficient light to see them. Occasionally a female enters the swarm to be seized immediately by a male. The pair then drops from the swarm and may settle to the ground; after the pair separates, the male rejoins the others in flight. Oviposition occurs as the female flies six to twelve inches above the water surface, touching the abdomen to the water at intervals. In ovipositing, the female does not rise and fall as in the mating flight but flies straight forward. The eggs are released a few at a time until the female is spent and she flies away to some nearby bush.

Taxonomy. Traver, 1933. *J. Elisha Mitchell Sci. Soc.* 48:173. Type species: *S. tripunctatum* (Banks); type locality: Milwaukee, Wisconsin. Classification of the species in this genus has been particularly difficult. There are several clusters of similar forms whose taxonomic interrelationships have not been well understood. Careful use of keys in the recent revision of the genus by Lewis (1974) should allow identification of most of the species as adults and nymphs. However, as Lewis clearly indicates, there remain a number of taxonomic problems in this genus. Flowers and Hilsenhoff (1975) have keyed nymphs and adults of the Wisconsin species.

Distribution. Nearctic and Neotropical. Mostly restricted to Nearctic, where all but one or two species occur east of the Rocky Mountains. The genus has been collected in the Neotropical realm as far south as Panama. Because of the extreme confusion over species identification in the past we have mostly recorded only the distribution of the species of *Stenonema* as given by Lewis. We believe, however, that the records of *S. terminatum* from British Columbia, eastern Oregon, Idaho, Wyoming, and Utah apply to an undescribed species related to *S. terminatum*. Another undescribed species occurs in Mexico and Central America.

Species	United States						Canada			Mexico		Central America
	SE	NE	C	SW	NW	A	E	C	W	N	S	
annexum Traver	X	...	X	...	...	...	...	...	...	...	...	...
ares Burks	...	...	X	...	...	...	...	...	...	...	...	...
bipunctatum (McDunnough)	...	X	X	...	...	...	X	X	...	...	...	...
carlsoni Lewis	X	...	...	...	...	...	...	...	...	...	...	...
exiguum Traver	X	...	X	...	...	...	...	...	...	...	...	...
femoratum (Say)	X	X	X	...	...	...	...	...	...	...	...	...
fuscum (Clemens)	...	X	X	...	...	...	X	X	...	...	...	...
integrum integrum (McDunnough)	X	X	X	...	...	...	...	...	...	...	...	...
integrum wabasha Daggy	...	...	X	...	...	...	...	...	...	...	...	...
ithaca (Clemens & Leonard)	X	X	...	...	...	...	X	...	...	...	...	...
lepton Burks	...	...	X	...	...	...	...	...	...	...	...	...
luteum (Clemens)	...	...	X	...	...	...	X	X	...	...	...	...
mediopunctatum (McDunnough)	...	X	X	...	...	...	X	X	...	...	...	...
modestum (Banks)	...	X	...	...	...	...	...	...	...	...	...	...
nepotellum (McDunnough)	X	X	X	...	...	...	X	...	...	...	...	...
placitum (Banks)	...	X	...	...	...	...	X	...	...	...	...	...
pudicum (Hagen)	X	X	...	...	...	...	...	...	...	...	...	...
pulchellum (Walsh)	...	X	X	...	...	...	X	...	...	...	...	...
quinquespinum Lewis	...	...	X	...	...	...	...	...	...	...	...	...
rubromaculatum (Clemens)	...	X	X	...	...	...	X	...	...	...	...	...
rubrum (McDunnough)	X	X	X	...	...	...	X	X	...	...	...	...
smithae Traver	X	...	...	...	...	...	...	...	...	...	...	...
terminatum (Walsh)	X	...	X	...	...	...	...	X	...	...	...	...
tripunctatum scitulum Traver	...	...	X	...	...	...	...	...	...	...	...	...
tripunctatum tripunctatum (Banks)	X	X	X	...	...	...	X	X	...	...	...	...
vicarium (Walker)	X	X	X	...	...	...	X	X	...	...	...	...
Species?	...	...	...	X	X	...	...	...	...	...	X	X

Subfamily Arthropleinae

This subfamily is based only on the unique genus *Arthroplea*. See the account of *Arthroplea* for distribution and characteristics.

Arthroplea Bengtsson (figs. 54, 56, 227, 238, 405)

Nymphal Characteristics. Body length: 7–10 mm. Body moderately flattened. Maxillary palpi more than six times as long as galea-lacinia, curved and with long setae on each side. Gill lamellae pointed at apex; abundant dark tracheation; fibrilliform portion absent. Three caudal filaments. Illustration: Fig. 405.

Nymphal Habitat. The nymphs inhabit ponds or channels with barely perceptible flow. They are frequently found on the stems of emergent sedges near the bottom. In a Massachusetts bog they were found to be abundant on the undersides of floating bog mats. The nymphs occupy the same habitat as do the univoltine northern *Aedes* mosquitoes.

Nymphal Habits. The nymphs move the gills rhythmically most of the time. When disturbed, the nymphs jump backward, aided by a strong stroke of the maxillary palpi. The nymphs are particle feeders and utilize both small animals and algae. The long setaceous maxillary palpi are drawn upward and backward, until they lie parallel to the lateral margins of the thorax and then are drawn forward, straining particles from the water. The particles caught in the setae of the palpi are removed by the setae of segment 2 of the labial palpi; the particles are then pushed up between the lingua and the superlingua to the mandibles. The mandibular canines may chop up large food particles; the molar areas grind and strain the food to concentrate it. The mouthparts are in continuous motion during feeding with a strong vortex reminiscent of the feeding of rotifers or some mosquito larvae.

Life History. Nymphal development appears to be complete in two months or less. Very small nymphs have been collected in mid-May, and emergence is complete by early June in Massachusetts. There is one generation per year.

Adult Characteristics. Wing length: 8 mm. Eyes of male separated dorsally by space 1⅓ width of median ocellus. Forelegs with basal tarsal segment slightly shorter than length of second segment; tarsi 2¼ to 2½ times length of tibiae. Hind wings with vein MA simple, unforked; vein Rs forms a regular fork. Genital forceps of male with three terminal segments (fig. 238).

Mating Flights. Solitary males have been seen flying over a meadow in late afternoon.

Taxonomy. Bengtsson, 1908. *Vet. Akad. Arsbok.* 6:239. Type species: *A. congener* Bengtsson; type locality: Varmland, Sweden. The only North American species is *A. bipunctata* McDunnough.

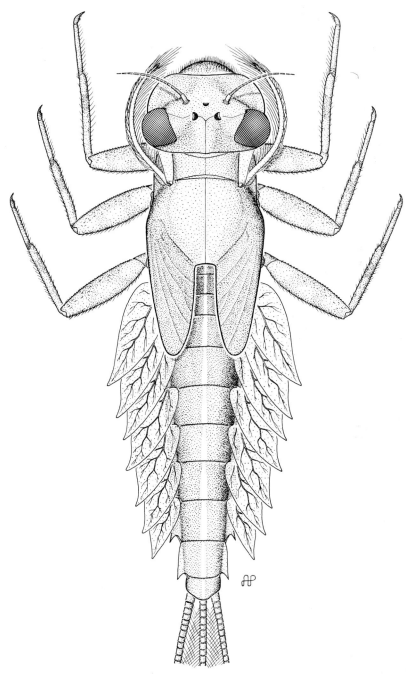

Fig. 405. *Arthroplea bipunctata*, nymph.

Distribution. Holarctic. In the Nearctic, eastern North America south and west to Massachusetts, Ohio, and Wisconsin.

Subfamily Pseudironinae

The subfamily contains only a single genus under which the characters and the distribution are given. The genus appears to be an isolated member of the family that originated early in the heptageniid lineage.

Pseudiron McDunnough (figs. 55, 234, 237, 370, 406)

Nymphal Characteristics. Body length: 11–15 mm. Head flattened, but area between eyes and antennae elevated on meson. Mouthparts adapted for predation; maxillary palpi three-segmented. Legs long and slender, directed posterolaterally; claws about as long as tibiae. Gill 1 with lamellae greatly reduced, lanceolate, with fibrilliform portion much longer; gills 2–7 elongate and slender; small fibrilliform tuft at base ventrally with long, narrow filament arising ventrally near center of lamellae. Three caudal filaments; terminal filament shorter and with dense fringe of setae on both sides; cerci with dense fringe on median side only. Illustration: Fig. 406.

Nymphal Habitat. The nymphs are found on sandy bottoms of medium-sized to large rivers.

Nymphal Habits. The nymphs lie on top of the sand, facing the current, with all three pairs of legs directed posteriorly and anchored in the sand. When disturbed, they are moderately strong swimmers, using their tails and bodies in an undulating motion. When seeking food, they actively move across the substrate; they are able to move laterally, forward, and backward equally well. They drift in the current during the night. Specimens have been maintained in the laboratory for several weeks by feeding them live chironomid larvae; they did not feed on dead larvae which were placed in the aquarium.

Life History. Adults have been collected as early as late April in Florida and as late as September in Utah.

Adult Characteristics. Wing length: 10–12 mm. Eyes of male large, separated by space 1¼ times width of median ocellus. Forelegs of male with tarsi twice length of tibiae; basal tarsal segment about ¾ length of second segment. Hind tarsi with four clearly differentiated segments; basal segment partially fused to tibia. Male genitalia illustrated in fig. 370.

Mating Flights. Not described.

Taxonomy. McDunnough, 1931. *Can. Entomol.* 63:91. Type species: *P. centralis* McDunnough; type locality: Lawrence, Kansas. The two described species of *Pseudiron* can be differentiated as adults. Nymphs and adults are rare in most collections and are poorly known.

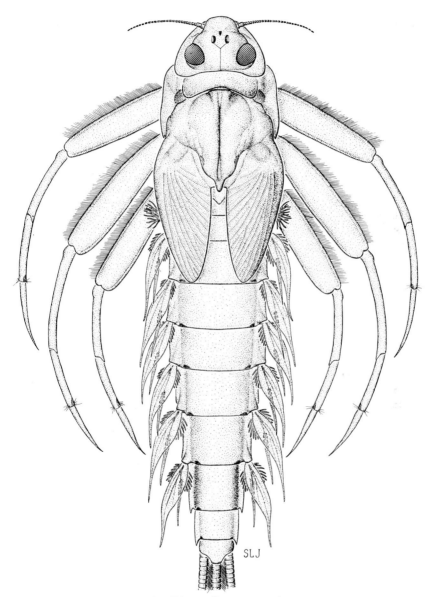

Fig. 406. *Pseudiron* sp., nymph.

Distribution. This relatively rare genus is restricted to North America and is known from scattered localities in central and southeastern North America west to Manitoba, Wyoming, and Utah.

Species	United States						Canada			Mexico		Central America
	SE	NE	C	SW	NW	A	E	C	W	N	S	
centralis												
McDunnough	X	...	X	...	...	...	...	X	...	...	...	...
meridionalis												
Traver	X	...	...	...	...	...	...	...	...	...	...	...
Species?	...	...	...	X	X	...	...	...	...	...	...	...

Subfamily Anepeorinae

The subfamily contains only the single rare genus *Anepeorus*, a highly adapted carnivore derived from advanced Heptageniinae. The subfamily is believed to be restricted to the Nearctic. The existing record from north-western China, based upon a female imago, must be considered as extremely questionable. See the generic account for characteristics and distribution in North America.

Anepeorus McDunnough (figs. 59, 239, 407)

Nymphal Characteristics. Body length: 6–8 mm. Body flattened. Head covered with dense setae; eyes small, located well within posterior and lateral margins of head; predatory mouthparts; mandibles with long, slender incisors, molar surfaces reduced. Legs flattened; dense fringe of long setae along posterior margins of femora and middle and hind tibiae; claws long and slender. Abdomen flattened and broad; terga densely covered with setae. Gills ventral, composed of slender lamellae and numerous filaments radiating out from a central plate. Three relatively bare caudal filaments. Illustrations: Fig. 407. (The nymph described is almost certainly the nymph of *Anepeorus*, but it has not been reared.)

Nymphal Habitat. Apparently only in large rivers. Nymphs have been collected by D. M. Lehmkuhl, University of Saskatchewan, in the North Saskatchewan and South Saskatchewan rivers in two to three feet of water from firmly compacted, rubble-laden substrate or loose gravel.

Nymphal Habits. The nymphs show preference for rocks over a sand substrate, and they move very rapidly. They are frequently collected in drift nets in certain large rivers. Active carnivores.

Life History. Unknown. Adults have been collected in June, July, and September.

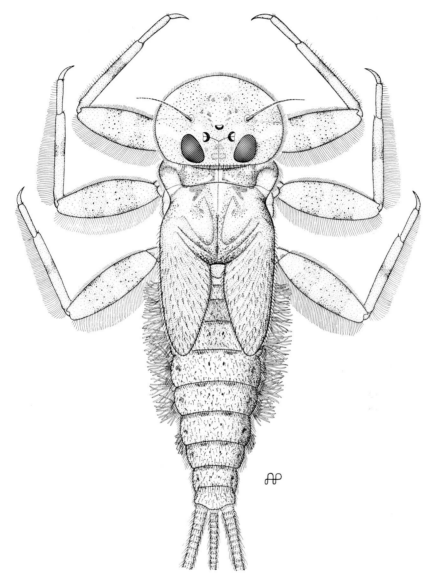

Fig. 407. *Anepeorus* sp., nymph.

Adult Characteristics. Wing length: 7–10 mm. Eyes of male widely separated dorsally. Forelegs only slightly longer than middle or hind legs; fore tarsi about ⅔ length of fore tibiae; basal tarsal segment ⅔ as long as segment 2. Basal costal crossveins of forewing weak; stigmatic crossveins slanted or partially anastomosed. Male genitalia illustrated in fig. 239.

Mating Flights. Unknown.

Taxonomy. McDunnough, 1925. *Can. Entomol.* 57:190. Type species: *A. rusticus* McDunnough; type locality: Saskatoon, Saskatchewan. Adults of this genus are very rare in collections, but specific identification should be possible.

Distribution. Nearctic, questionably Oriental. In the Nearctic, scattered locality records of *A. simplex* (Walsh) from Illinois, Iowa, and Georgia; *A. rusticus* McDunnough from Saskatchewan and Utah.

Subfamily Spinadinae

Only a single genus with one species is known. The characters of the genus are those of the subfamily. The phylogenetic relationship of this subfamily within the family is obscure.

Spinadis Edmunds and Jensen (fig. 408)

Nymphal Characteristics. Body length: 9 mm. Body somewhat flattened; front of head evenly convex; two pairs of tubercles between eyes; eyes of female extend to or near posterolateral angle of head; eyes of male extend beyond posterolateral angle; predatory mouthparts; mandibles with long, slender incisors, molar surface reduced. Thorax with a pair of tubercles dorsally on each segment. Legs long and slender; a digitlike projection at apex of each femur; claws long and slender, a single denticle near base of each. Abdomen somewhat flattened. Gills on segment 1 lateroventral, on segments 2 and 3 ventral; lamellae on gills 1 to 3 reduced to slender lanceolate structures, fibrilliform portion large; gills on segments 4 to 7 progressively more lateral with progressively broader lamellae. Two caudal filaments only, with long setae on mesal sides in basal ¾ of filaments. Illustration: Fig. 408.

Nymphal Habitat. Known only from large rivers. Nymphs have been taken in drift nets in water over ten feet deep, collected from the bottom in water about five feet deep after the river had dropped in volume, and from a sunken log in two feet of water.

Nymphal Habits. Almost unknown. The larvae feed on chironomids.

Life History. Unknown. The adults probably emerge in late June or early July.

Adult Characteristics. Unknown.

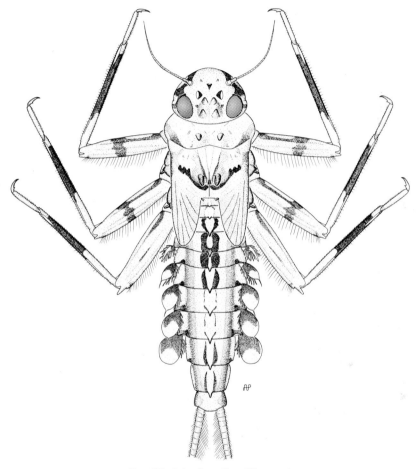

Fig. 408. *Spinadis wallacei*(?), nymph.

Mating Flights. Unknown.

Taxonomy. Edmunds and Jensen, 1974. *Proc. Entomol. Soc. Wash.* 76:456. Type species: *S. wallacei* Edmunds and Jensen; type locality: Altamaha River, Georgia. It is uncertain whether there is only one species in the three known collection areas. Figure 408 shows an Indiana specimen.

Distribution. Nearctic. Georgia, Indiana, and Wisconsin.

Family LEPTOPHLEBIIDAE

This family is placed in the Leptophlebioidea. The family is relatively isolated and retains a number of primitive characteristics. It shares a number of characters with Ephemerellidae and Tricorythidae and also with the Ephemeroidea. The pre-leptophlebiid mayflies may have given rise to all of the families except those assigned to the Heptagenioidea. For this reason, the decision about which other families should be in the Leptophlebioidea, if any, is not an easy one to resolve to the satisfaction of most workers. The Leptophlebiidae seem to occupy a postgroup relationship to the ancestral form which also gave rise to the Ephemerellidae and a pregroup relationship to the Ephemeroidea.

The family is extremely widespread, reaching its maximum diversity in the southern hemisphere. In Australia, New Zealand, South America, and Africa it forms the principal component of the mayfly fauna. The family reaches a few oceanic islands.

The North American fauna has a variety of genera, some of Holarctic distribution and others of Neotropical affinity. Some of the family characteristics given below do not apply to certain extralimital genera.

Nymphal Characteristics. Body more or less depressed; head hypognathous or prognathous. If body and head strongly depressed, upper surface of head partially formed from mandibles. Maxillary and labial palpi three-segmented. Abdominal gills on segments 1–7 or 1–6; gills forked or of two plates, at least on segments 2–6, these appearing as a single tuft in *Habrophlebia*; platelike gills with acute apex on projection or projections. Three caudal filaments, with more or less sparse whorls of setae at apex of each segment.

Adult Characteristics. Eyes of male distinctly divided into upper portion of large facets and lower portion with smaller facets. Wings without short free marginal intercalaries between longer veins; two to four long intercalaries between veins CuA and CuP; vein CuP usually rather strongly recurved. Hind wings present (except in the lower austral genus *Hagenulopsis*). Three caudal filaments.

Choroterpes Eaton (figs. 144–146, 330, 331, 377, 379, 409)

Nymphal Characteristics. Body length: 5–8 mm. Body flattened. Head prognathous; labrum widened and deeply emarginate. Claws with row of minute blunt denticles. Gills somewhat variable; in all species in Canada and United States (except one species in Texas), gills on segment 1 single and unbranched, linearly lanceolate; gills 2–7 double, lamelliform, each lamella having a lanceolate terminal extension between two pointed lateral projections; in some nymphs from Arizona gills on segments 2–7 with three subequal slender pointed projections; in some nymphs from Texas and Mexico gills on segment 1 asymmetrically forked and gills on segments 2–7 with three somewhat subequal lanceolate terminal extensions (see **Taxonomy**). Illustration: Fig. 409.

Nymphal Habitat. The nymphs are found in crevices on the underside of rocks, logs, or sticks anchored in streams and also among accumulations of leaves intermixed with silt or on the underside of algal mats in slow-flowing portions of streams. They also occur on the margins of lakes where there is some wave action, occupying crevices on the underside of firmly anchored objects.

Nymphal Habits. Negatively phototropic and strongly thigmotactic. The gills are held above and close to the abdomen and are frequently vibrated. The nymphs appear to be herbivorous.

Life History. The time required for nymphal development is unknown; development may take from six months to one year, depending on the part of the country in which the species occurs. The transformation from nymph to adult occurs after dark. When ready to emerge, the nymph swims vigorously to the surface, and the subimago bursts free, floats for a moment on the exuviae, then flies to a nearby support. The subimago lives from eight to ten hours. In Florida emergence occurs throughout the year, reaching a peak during spring and early summer. Farther north emergence occurs from July to early September. *C. oklahoma* adults were collected on March 20.

Adult Characteristics. Wing length: 5–9 mm. Bodies of both sexes blackish or brownish; abdominal venter much paler. Wings with anterior longitudinal veins brownish; more posterior ones colorless as are the crossveins. Hind wings small with prominent costal projection near midpoint of length; subcosta reaches wing margin just beyond costal projection. Male genitalia illustrated in fig. 379.

Mating Flights. Unknown.

Taxonomy. Eaton, 1881. *Entomol. Mon. Mag.* 17:194. Type species: *C. picteti* (Eaton); type locality: Geneva, Switzerland. It should be possible to identify males of *Choroterpes* to species; however, not enough is known about the nymphs to provide a satisfactory key. Kilgore and Allen (1973) have described the nymphs of several Mexican and western North American species. Allen (1974) has divided the genus into two subgenera, but we have not revised the keys and generic accounts to include this taxonomic change. The species placed in the subgenus *Neochoroterpes* are *crocatus*, *kossi*, and *mexicanus*. In some individuals of some species of *Neochoroterpes*, the first

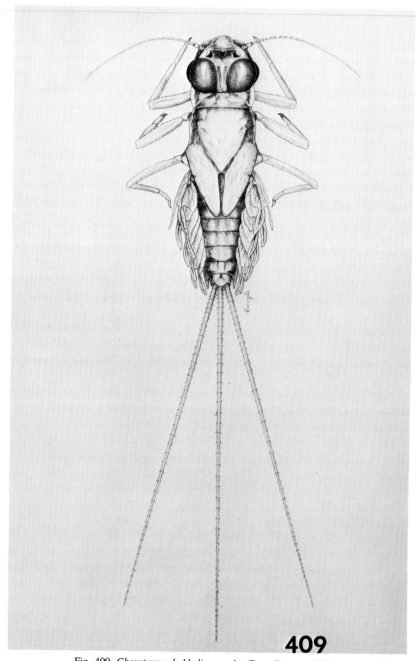

409

Fig. 409. *Choroterpes hubbeli*, nymph. (From Berner, 1950.)

gill is asymmetrically forked (fig. 146a). Gills 2–7 bear three lanceolate terminal projections (fig. 146b) in most species, but the three projections are slender and acuminate in *C. kossi* (see fig. 145). Edmunds collected nymphs and cast exuviae of this or a closely related species in Arizona and at the same time and place collected adults tentatively identified as *C. inornata*. Kilgore and Allen (1973) place a typical nymph of the subgenus *Choroterpes* as the nymph of *C. inornata*. It is interesting that the gills of *C. kossi* resemble the gills of the Old World subgenus *Euthraulus*. The subgenus *Neochoroterpes* can be identified as nymphs by the presence of gills of the types in figs. 145 and 146; the gills of the subgenus *Choroterpes* are similar to those in fig. 144. The adult hind wings of the one known species of the subgenus *Neochoroterpes* (fig. 331) differ from those of the subgenus *Choroterpes* (figs. 330b, c). It is not known whether this character will distinguish other *Neochoroterpes* adults. The nymph of *C. kossi* certainly differs in gill type from the nymphs of other *Neochoroterpes*. It appears that the phyletic and taxonomic relationships of the species of *Choroterpes* in Mexico and north to Texas and Arizona are far from resolved but are of considerable biogeographic interest.

Distribution. Ethiopian, Oriental, Holarctic, and Neotropical. Widespread in North and Central America. The nymphs of *Choroterpes* from Mexico, Texas, and Arizona suggest the possibility of two groups of *Choroterpes*, one of Holarctic and one of Neotropical affinities.

Species	United States						Canada			Mexico		Central America
	SE	NE	C	SW	NW	A	E	C	W	N	S	
albiannulata McDunnough	...	...	...	X	X	...	...	...	X	...	...	...
atramentum Traver	...	...	...	...	...	...	...	...	...	...	...	X
basalis (Banks)	...	X	X	...	...	...	X	...	...	...	...	...
crocatus Allen	...	...	X	...	...	...	...	...	...	...	...	...
ferruginea Traver	...	X	...	...	...	...	...	...	...	...	...	...
fusca Spieth	...	X	...	...	...	...	X	...	...	...	...	...
hubbelli Berner	X	...	...	...	...	...	...	...	...	...	...	...
inornata Eaton	...	...	...	X	...	...	...	...	...	X	...	...
kossi Allen	...	...	...	X	...	...	...	...	...	...	...	...
mexicanus Allen	...	...	X	...	...	...	...	...	...	X	X	...
nanita Traver	...	...	X	...	...	...	...	...	...	...	...	...
nervosa Eaton	...	...	...	...	...	...	...	...	...	...	...	X
oaxacaensis Brusca & Allen	...	...	...	...	...	...	...	...	...	X	...	...
oklahoma Traver	...	...	X	...	...	...	...	...	...	...	...	...
terratoma Seemann	...	...	...	X	...	...	...	...	...	...	...	...
vinculum Traver	...	...	...	...	...	...	...	...	...	...	...	X

Habrophlebia Eaton (figs. 143, 303, 306)

Nymphal Characteristics. Body length: 4–5 mm. Body slender, only slightly flattened. Head more or less hypognathous; eyes lateral; claws relatively short with a single ventral row of denticles. Gills on abdominal segments 1–7 double and narrowly lamelliform at base, divided near base into a number of linear filaments outspread at sides of abdomen. Illustration: Eaton, 1883–88, pl. 36.

Nymphal Habitat. In small streams near the edges among vegetation or in leaf debris where there is some accumulation of silt.

Nymphal Habits. Not described.

Life History. Not determined. Probably requires from eight to twelve months to complete development, depending on the part of the country in which the species occurs. Adults have been collected from April to August.

Adult Characteristics. Wing length: 4.5–5.0 mm. Both sexes brown. Forelegs of male longer than body; tarsi ⅔ length of tibiae. Tibiae of all legs in both sexes longer than femora. Forewings with few crossveins in the stigmatic area; all longitudinal veins are hyaline or tinted basally only; crossveins almost invisible. Hind wings small with sharp costal projection near midpoint of length; subcosta reaches wing margin between costal projection and wing apex. Male genitalia illustrated in fig. 306.

Mating Flights. At Ithaca, New York, adults of one species fly in the afternoon in small compact swarms in forest openings beside streams.

Taxonomy. Eaton, 1881. *Entomol. Mon. Mag.* 17:195. Type species: *H. fusca* (Curtis); type locality: Great Britain.

Distribution. Holarctic; in the Nearctic, in eastern United States and Canada south to Florida. The single species, *H. vibrans* Needham, is now the only species of the genus in North America.

Habrophlebiodes Ulmer (figs. 148, 155, 157, 304, 307, 410)

Nymphal Characteristics. Body length: 4–6 mm. Body slender, somewhat flattened. Head more or less hypognathous, somewhat flattened; eyes lateral; anterior margin of labrum rather deeply cleft. Claws long and slender, each bearing a single ventral row of denticles. Gills on abdominal segments 1–7 similar; each gill has slender base which subdivides into two long, slender, lanceolate filaments; a single multibranched trachea present in each lamella. Illustration: Fig. 410.

Nymphal Habitat. In slow to moderately swift streams where the nymphs may occur in leaf debris. Although the nymphs are sometimes found in riffles, they are most common among the exposed roots of terrestrial plants, in submerged vegetation along the banks of streams, and in crevices of water-

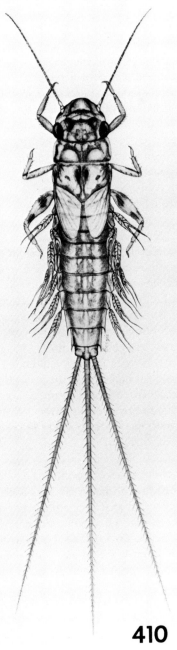

410

Fig. 410. *Habrophlebiodes brunneipennis*, nymph. (From Berner, 1950.)

soaked and partially rotted logs. Early instars often occur among the pebbles in riffles, but as the nymphs mature they migrate into quieter water. The streams inhabited by nymphs usually drain heavily wooded areas or bogs and are somewhat acid and tinged with brown.

Nymphal Habits. The gills are kept in constant motion. When the nymphs are at rest, the gills are held out stiffly from the body at about a 45° angle and are slowly waved forward and backward. While swimming, the nymphs press the gills against the abdomen. They swim awkwardly with undulatory movements, and they progress most efficiently by crawling. The caudal filaments help very little in propulsion. Strongly negatively phototropic and positively rheotropic.

Life History. The time required for nymphal development has not been determined, but development probably takes from six to twelve months, depending on the climate. Emergence usually occurs in morning or early afternoon. Before transforming, the nymph crawls out of the water, sometimes as far as an inch above the surface. After a few moments the subimago appears, rests a moment, then flies to a nearby support. The subimago lives for twelve to fourteen hours. In Florida emergence occurs throughout the year; farther north emergence is restricted to the late spring and summer months.

Adult Characteristics. Wing length: 4.5–5.0 mm. Slender, brownish mayflies. Fore tarsi of male as long as fore tibiae. Wings colorless or tinged with brown; longitudinal veins well marked with brown, but crossveins may be colorless. Hind wings with thumb-shaped costal projection beyond midpoint of length; subcosta ends just beyond projection. Female with small ovipositor. Male genitalia illustrated in fig. 307.

Mating Flights. Swarms of five to fifteen males fly only a few inches above the water surface; larger swarms of up to a hundred males fly at heights of two to five feet. The rise and fall of individual insects is about one foot. Swarming occurs in sunlit areas over the water where there is a break in the vegetation. The female flies into the swarm of males and rises and falls with the swarm until approached from below and seized for copulation by a male. The mating pair then drifts out of the swarm. Oviposition has not been reported.

Taxonomy. Ulmer, 1920b. *Archiv für Naturgeschichte* 85:39. Type species: *H. americana* (Banks); type locality: Passaic, New Jersey. The adults of this genus should be identifiable to species. No keys to the nymphs are available. The nymphs of this genus very closely resemble the nymphs of some species of *Paraleptophlebia*. Berner (1975) has reviewed the southeastern species of *Habrophlebiodes*.

Distribution. Oriental and Nearctic. In the Nearctic, in central and eastern North America south to Florida, Illinois, and Arkansas.

Species	United States						Canada			Mexico		Central America
	SE	NE	C	SW	NW	A	E	C	W	N	S	
americana (Banks)	X	X	X	...	...	...	X	...	...	...	...	...
annulata Traver	...	...	X	...	...	...	...	...	...	...	...	...
brunneipennis Berner	X	...	...	...	...	...	...	...	...	...	...	...
celeteria Berner	X	...	...	...	...	...	...	...	...	...	...	...

Hagenulopsis Ulmer (figs. 136, 153, 160, 164, 165, 212, 213, 218, 376)

Nymphal Characteristics. Body length: 4.2–6.6 mm. Body flattened. Head prognathous. Apical segment of labial palpi narrowed distally; glossae with spatulate hairs. Claws with a row of small denticles plus a large apical denticle or with a row of denticles increasing in size toward the apex. Hind wing pads absent. Abdominal gills deeply forked, each branch long and slender; gills of segment 1 similar in shape to those of other segments (gills similar to those of some *Paraleptophlebia*, but the two genera are found in separate ranges). Illustration: Parts in Traver, 1943.
Nymphal Habitat. Not described.
Nymphal Habits. Unknown.
Life History. Unknown.
Adult Characteristics. Wing length: 6 mm. Eyes of male turbinate, covering much of dorsum of head. Fore tarsi of male about ⅔ as long as tibiae. Hind wings absent. Female with an ovipositor. Male genitalia illustrated in fig. 376.
Mating Flights. Unknown.
Taxonomy. Ulmer, 1920b. *Archiv für Naturgeschichte* 85:34. Type species: *H. diptera* Ulmer; type locality: Humboldt District, Isabella Region, Santa Catarina, Brazil. Nothing is known of the taxonomy of the species in Central America.
Distribution. Neotropical. A tropical genus with nymphs of one undetermined species known from Honduras.

Hermanella Needham and Murphy (figs. 140, 310)

Nymphal Characteristics. Body length: 4–11 mm. Body flattened. Head rectangular, prognathous; eyes dorsal, labrum as wide as head; maxillary palpi lie along side of head, visible dorsally. Wing pads of the type species appear almost to cover the much shortened abdomen. In Central American species, wing pads of usual size. Gills of two lamellae present on segments 1–6 or 1–7; apex of each lamella with two lateral rounded lobes and a short median fingerlike projection. Illustration: Demoulin, 1955, fig. 4.

Nymphal Habitat. The nymphs are found in streams, where they crawl among the stones to escape the full force of the current. They are typically found in well-aerated water flowing over rocky bottoms.

Nymphal Habits. Not described.

Life History. Unknown.

Adult Characteristics. Wing length: 4–10 mm. Eyes of male large, nearly contiguous dorsally. Claws dissimilar on all legs. Wings similar to those of *Traverella*. Hind wings with acute costal projection. Male genitalia illustrated in fig. 310; a pair of submedian stout projections arises between forceps bases.

Mating Flights. Unknown.

Taxonomy. Needham and Murphy, 1924. *Bull. Lloyd Lib*. 24, Entomol. Ser. 4:39–40. Type species: *H. thelma* Needham and Murphy; type locality: Iguazú Falls, Argentina. There appear to be several species as seen in nymphs from Honduras; adults from the area have not been described. The correct association of the nymphs of *Hermanella* with the adult forms was made through specimens reared by Dr. and Mrs. W. L. Peters, who kindly provided the unpublished information. The Central American nymphs differ from those of *H. thelma* in a few details, especially in having a relatively smaller head and a longer abdomen and in their larger overall size. The long wing pads as illustrated for the type by Needham and Murphy (1924) appear to be atypical of the species. Dr. and Mrs. Peters have not found such wing pads on individuals believed to be of this species from the type locality in Argentina. (Peters is currently working on the taxonomic problems of *Hermanella* and its allies.)

Distribution. Neotropical, from extreme southern Brazil and northern Argentina north to Honduras. One or more undetermined species are found in Central America.

Hermanellopsis Demoulin (figs. 163, 308, 332)

Nymphal Characteristics. Body length: 5 mm. Head prognathous. Apical segment of labial and maxillary palpi conical, less than ½ length of segment 2. Hind wing pads present. Gills deeply forked, each branch long and slender; gills on segment 1 similar in shape to those of other segments; gills largest on segments 3–5. (The gills are similar to those of some *Paraleptophlebia*, but the two genera are found in separate ranges.)

Nymphal Habitat. One nymph was collected from a small brook in the highlands of Panama. The stream was clear and about eight feet wide.

Nymphal Habits. Unknown.

Adult Characteristics. Wing length: 5 mm. Eyes of male divided into an upper orange portion and a lower gray portion about ½ as large, upper parts meet on vertex. Wings long and narrow, about ⅓ as wide as long; cubital area with two intercalaries. Hind wings pointed apically, with an acute costal projection near midpoint of length; vein Sc ends just beyond projection. Female without ovipositor. Male genitalia illustrated in fig. 308.

Mating Flights. Unknown.

Taxonomy. Demoulin, 1955. *Bull. Inst. Roy. Sci. Natur. Belg.* 31(20):8. Type species: *H. incertans* (Spieth). The taxonomy of this genus is poorly known, and the identity of the species in Panama is in question. The adults of *incertans* were described by Spieth (1943) as being *Hermanella*, a genus described by Needham and Murphy (1924) from nymphs only of the species *H. thelma*. Traver (1947a) noted that the venation visible in the hind wing pads of *Hermanella thelma* nymphs differed from that in the hind wings of *incertans* adults. On the basis of Traver's comment Demoulin (1955) established a new subgenus, *Hermanellopsis*, and placed *incertans* Spieth as the type species. It is now certain that the species *thelma* and *incertans* belong neither to the same genus nor to allied genera. (See the description of *Hermanella* for adults of that genus.)

The nymphs of what we herein regard as the genus *Hermanellopsis* at this stage of knowledge are almost identical with the nymphs that Traver (1943) described as belonging to a *Hagenulopsis* ally. Peters (1969) reports that the nymphs described by Traver are indeed *Hagenulopsis*. We have studied a single nymph from Panama whose hind wing pads show clearly a wing venation of the *incertans* type and which therefore must be regarded either as a nymph of *H. incertans* or more probably of a related species. The genus *Hermanellopsis* is very closely related to *Hagenulopsis*, as evidenced by details of the nymphal gills, claws, legs, and mouthparts. In fact, it is distinguished from *Hagenulopsis* only by the presence of hind wing pads and the fact that the abdomen has posterolateral projections only on segments 8–9 rather than on segments 6–9 as in *Hagenulopsis*. This latter character may prove to be of specific rather than generic significance.

The real taxonomic situation is probably more complex than is indicated by the present fragmentary knowledge. Although *Hermanellopsis* has hind wings and *Hagenulopsis* does not, the overreliance on the loss of the hind wings as a taxonomic character may result in generic groupings that are not natural. Much more rearing of nymphs is required in this complex. The gills of *Hagenulopsis* and *Hermanellopsis* are similar to those of the nymphs of some *Paraleptophlebia*, but the first two genera are Neotropical while *Paraleptophlebia* is Holarctic — thus *Paraleptophlebia* is geographically separated from *Hermanellopsis* and *Hagenulopsis* by Mexico. The similarities of *Paraleptophlebia* are only superficial, and the setae of the legs, the denticles of the claws, and the mouthparts will give a clear-cut identification if nymphs of *Paraleptophlebia, Hermanellopsis*, or *Hagenulopsis* are found in intervening localities in Mexico.

Distribution. Northern South America (Guiana area) north to Chiriquí Province, Panama.

Homothraulus Demoulin (figs. 137, 154, 166, 334, 375)

Nymphal Characteristics. Body length: 5–9 mm. Body somewhat flattened. Head prognathous. Labrum about ⅔ width of head, slightly wider than clypeus, sides rounded. Legs with only scattered setae. Claws with a prominent pigmented subapical denticle and numerous small pale denticles. Gills forked, slender (about 1/6 to ⅛ as wide as long), gills 3 to 4 times as long as the segment from which each arises. Posterolateral projections on segments 8 and 9. Illustrations: Parts in Traver, 1960. (Nymph figured by Demoulin, 1955, is not *Homothraulus*.)

Nymphal Habitat. In Uruguay specimens were taken from a shallow, rapid river having a bottom of loose boulders and sand in some places but having deep pools with slow currents and sand or silt bottoms elsewhere. The specimens from Guatemala were taken in a stream one inch deep and four inches wide at 10,000 feet elevation. The Texas nymphs were found with the nymphs of *Thraulodes gonzalesi* in a creek.

Nymphal Habits. Unknown.

Life History. Unknown.

Adult Characteristics. Wing length: 5.5–8.0 mm. Eyes of male contiguous dorsally. Costal projection of hind wings well developed, acute, near midpoint of length of wing. Vein MP of hind wings simple. Genitalia of male as in fig. 375.

Mating Flights. Flights were observed in the morning in Uruguay.

Taxonomy. Demoulin, 1955. *Bull. Inst. Roy. Sci. Natur. Belg.* 31(20):11. Type species: *H. misionensis* (Esben-Peterson); type locality: Bompland, Misiones Province, Argentina. The discovery of nymphs in Texas that seem to be assignable to this genus prompted us to include *Homothraulus* in this book, although the taxonomic problems were solved so late, albeit far from satisfactorily, that the complex problem of adding new couplets to the nymphal and adult keys was not undertaken. Presumably *Homothraulus* occurs only in limited areas, and determination is not likely to be a major problem. The adult and nymphal characters were drawn only from the segregate of *Homothraulus* from Argentina, Central America, and Texas described in the next paragraph.

The nymphs key down to a group of three genera which we have separated first on distribution and secondarily on structure. There can be confusion of *Homothraulus* with *Thraulodes* because the labrum of *Homothraulus* is about two-thirds as wide as the head; however, the labrum of *Homothraulus* is only slightly wider than the clypeus, but in *Thraulodes* it is clearly much wider. Furthermore, *Thraulodes* nymphs have a dense row of setae on the posterior margins of the femora and tibiae. In Texas *Homothraulus* also could be confused with species of *Paraleptophlebia* that have narrow gills without pigmented tracheal branches (if such species occur in Texas).* The claws of *Homothraulus*, with a prominent pigmented subapical denticle and fine pale denticles basally, will easily separate these nymphs from *Paraleptophlebia* nymphs, whose claws have denticles that form a slender series. *Paraleptophlebia* is not closely related to *Homothraulus*, *Hermanellopsis*, and *Hagenulopsis*. In *Paraleptophlebia* nymphs the mouthparts are directed downward (hypognathous) as illustrated in figure 412, whereas in the nymphs of the other genera the mouthparts tend to be directed forward (prognathous), similar to those of *Thraulodes* in figure 413. In Central America the distinctive claws of *Homothraulus* nymphs will probably distinguish them from the other two genera; the large prominent pigmented subapical denticle contrasted with the fine pale denticles basally seem distinct from the claws of both *Hermanellopsis* and *Hagenulopsis*. The Central American *Homothraulus* nymphs are up to 8 mm long when mature, and those of *Hagenulopsis* and *Hermanellopsis* are generally much smaller. Nevertheless, there is so little known of these genera in Central America that difficulties in taxonomy are to be expected.

Homothraulus adults follow the key through to a couplet that leads to *Traverella* and *Hermanella*. The hind wings of *Homothraulus* have a more prominent costal projection and fewer veins in the hind wings than do *Traverella* and *Hermanella*, and the male genitalia lack the extra projections found in the other two genera.

Our placement of the specimens from Texas, Central America, and Argentina in *Homothraulus* is not very satisfactory without isolating the group as a separate subgenus (or possibly placing it as a separate genus). We are reluctant to do so without a more comprehensive study of these and some apparently related forms. The only species that has been reared is an Argentina species (apparently undescribed).

We have no reared or associated specimens of *Homothraulus* from North or Central America. In 1972 we received nymphs from Guatemala which we set aside as being "near *Homothraulus*." In 1974 Michael S. Peters brought unidentified leptophlebiid nymphs from Caldwell County, Texas, to the University of Utah. Many years ago two adult male Leptophlebiidae from Dallas County, Texas, were submitted to Edmunds for identification. The males were identified as a probable new genus and returned at the request of the

---

* The only *Paraleptophlebia* known to occur in Texas is the one whose gills are shown in figure 152. This nymph has gills like *bradleyi*, which Berner (1975) has transferred from *Paraleptophlebia* to *Leptophlebia*. The confusion of *Homothraulus* with *Paraleptophlebia* or *Leptophlebia* in Texas seems increasingly unlikely.

sender; they subsequently disappeared and Edmunds was left only with recol-
lections of the characters. The adults reared from nymphs in Argentina by
Luis Peña in 1967 had penes that immediately reminded Edmunds of the
Texas adults, and the Texas nymphs which are congeneric with the Argentina
nymphs seem to complete the association. In 1975 we found adults from a
separate locality in Guatemala that have penes and other features similar to the
reared specimens from Argentina.

It now appears that a distinct assemblage of species occurs in scattered
localities in Texas, Guatemala, and Argentina. The nymphs have much narrower gills than do the typical *Homothraulus* nymphs described by Traver
(1960). In the hind wings the costal projection is more acute and the forks of
the wing veins are less complete in the hind wings than in typical *Homothraulus*. The penes of the males from Argentina, Guatemala, and Texas (as
the Texas specimens are recalled by Edmunds) have large, laterally directed
appendages whereas typical *Homothraulus* males have small appendages directed anteriorly. *"Thraulus"* *caribbeanus* appears to represent a third segregate of *Homothraulus* in a broad sense.

Distribution. Argentina and Uruguay, north to Texas. The typical form of
Homothraulus does not occur in Central or North America. The genus is not
known from Mexico despite the large collections of R. K. Allen, most of
which have been studied. Almost certainly the genus will be found there in
some habitats.

Leptophlebia Westwood (figs. 138, 147, 289, 299, 411)

Nymphal Characteristics. Body length: 7–15 mm. Body only slightly flattened. Head more or less hypognathous. Claws with a uniform row of ventral
denticles. Gills on abdominal segments 2–7 double and lamelliform, each
lamella terminating in a single acute filament; gills on segment 1 slender,
forked. Illustration: Fig. 411.

Nymphal Habitat. In parts of the range the nymphs develop in ponds, lakes,
or quiet eddies along the banks of streams. In the southeast they are found
only in flowing water or in backwaters of streams. They are poorly adapted for
swift water and are almost always found in quiet portions of streams or in
pools recently cut off from streams. They prefer quiet areas near the banks,
pools where dead leaves accumulate, and submerged mossy banks where they
crawl amid the bases of the plants. Before emerging the nymphs migrate into
shallow marginal waters or tributaries where they become highly concentrated
in favorable situations. In Manitoba *L. cupida* has been shown to migrate up
temporary streams formed by melting snow, with some nymphs even leaving
the water and crawling along the banks as they move upstream; the nymphs
migrate as much as four hundred yards a day. The large gill expanse is well

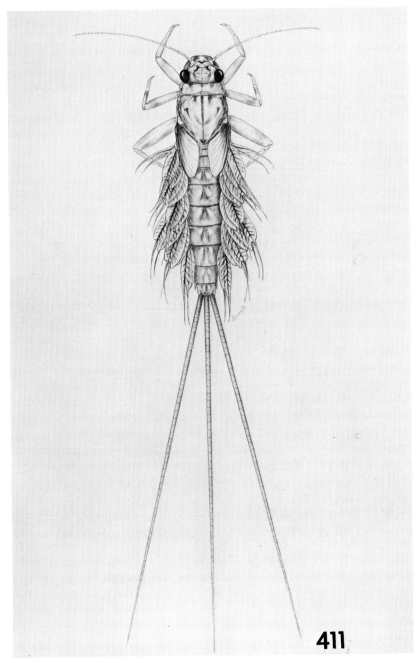

411

Fig. 411. *Leptophlebia intermedia*, nymph. (From Berner, 1950.)

adapted for the low oxygen content of the stagnant water in which the nymphs may live. Mature nymphs have been found in the northern part of the range in isolated small pools fed by springs and have been observed emerging from water at 9° C (48° F).

Nymphal Habits. Omnivorous, feeding on detritus, diatoms, algae, and cast skins of nymphs of the same or other species. The nymphs are negatively phototropic, seeking the dark underside of any materials that are available. At night they move to the upper surface of the substrate, but they tend to hide during the day. Nymphs about to emerge undergo a phototactic reversal and crawl to the upper side of leaves or onto sticks, logs, or any other available support near the surface of the water.

Life History. The time required for nymphal development is less than a year. Some nymphs crawl out of the water to emerge, but generally emergence takes place at the surface. Emergence may occur at any time of day with maximum transformation in early afternoon. The subimaginal stage lasts from eighteen to twenty-nine hours. In Florida emergence begins as early as late January with the peak in February and early March. Farther north emergence occurs as early as February and continues on into May. In southern Canada species emerge in late May and June, and *L. cupida* has been reported emerging in Michigan as late as August 31.

In the Bigoray River, Alberta, Clifford (1969, 1970) reports that there is one generation of *L. cupida* each year. Nymphs of the new generation first appear in large numbers in August and September. They grow rapidly during the remainder of the ice-free season but very little in winter, which extends from November through most of April. There is a continuous influx of small nymphs into the population through December, indicating delayed hatching. In late April or early May, when the ice breaks up, most of the nymphs move from the main stream into small, usually vernal, tributaries and then into the marshy regions drained by the tributaries. Most of the nymphs transform and emerge from the marshes in May and June. The female imagoes fly back to the main stream to oviposit.

Adult Characteristics. Wing length: 5.5–13.0 mm. Bodies predominantly dark yellow-brown. Fore tarsi of male from 1 to 1⅔ times as long as tibiae. Wing color varies from clear or partly stained with brown to fuscous; all veins and most crossveins brown. Hind wings without costal projection. Male genitalia illustrated in fig. 289. Terminal filament either equal to cerci in length and thickness or shorter and weaker than cerci.

Mating Flights. Mating flights usually take place over water at heights ranging from two or three feet to as high as the tops of trees bordering the stream; swarms may form in sunny forest openings some yards from the stream. The flight occurs from midafternoon to dark. The female oviposits by repeatedly touching the tip of the abdomen to the water surface while flying just above the water. Swarms of *L. cupida* appear to be small, with no more than twelve insects performing simultaneously.

Taxonomy. Westwood, 1840. *Introduction to a Modern Classification of Insects* (2 vols.), 2:31. Type species: *L. marginata* (Linnaeus); type locality: Europe. Adults should generally be identifiable to species, but nymphs cannot be determined to species in most cases.

The separation of the genera *Leptophlebia* and *Paraleptophlebia* is simple with one exception. The nymphs are easily distinguished by gill form, and the penes of adult males of *Leptophlebia* have a relatively uniform shape not found among the diverse types in *Paraleptophlebia*. The females of all *Leptophlebia* except *L. johnsoni* have the terminal filament shorter and thinner than the cerci, while the caudal filaments are subequal in *Paraleptophlebia*.

We have inserted an extra couplet in the key to adults for placing *Leptophlebia johnsoni* to genus on the basis of superficial characters and distribution. Burks (1953) has used the wing venation to separate the two genera, but in some species this character is very difficult to use and appears to be invalid. The wings of *Leptophlebia* (fig. 299) and *Paraleptophlebia* (fig. 300) show numerous differences in our drawings, but these differences relate mostly to size. The large species of both genera have more crossveins, more marginal veinlets, larger hind wings, and so on. Unfortunately, *Leptophlebia johnsoni* is the smallest North American *Leptophlebia*, and its wing venation does not appear to differ in any character from that of the larger species of *Paraleptophlebia*. Within the geographic range of *Leptophlebia johnsoni* most *Paraleptophlebia* are smaller than *Leptophlebia johnsoni*, and the caudal filaments of most are uniformly tan or uniformly pale rather than tan with reddish brown rings at the joinings as found in *L. johnsoni*.

Distribution. Holarctic. In the Nearctic, most species occur in central and eastern North America south to Illinois, Ohio, and Florida; in western North America the genus occurs south to Oregon, Utah, and Colorado.

Species	United States						Canada			Mexico		Central America
	SE	NE	C	SW	NW	A	E	C	W	N	S	
austrina (Traver)	X	. . .	. . .	. . .	. . .	. . .	. . .	. . .	. . .	. . .	. . .	. . .
collina (Traver)	X	. . .	. . .	. . .	. . .	. . .	. . .	. . .	. . .	. . .	. . .	. . .
cupida (Say)	X	X	X	. . .	. . .	. . .	X	X	X	. . .	. . .	. . .
grandis (Traver)	X	. . .	. . .	. . .	. . .	. . .	. . .	. . .	. . .	. . .	. . .	. . .
gravastella (Eaton)	. . .	. . .	. . .	X	X	. . .	. . .	. . .	. . .	. . .	. . .	. . .
intermedia (Traver)	X	. . .	. . .	. . .	. . .	. . .	. . .	. . .	. . .	. . .	. . .	. . .
johnsoni McDunnough	X	X	. . .	. . .	. . .	. . .	X	. . .	. . .	. . .	. . .	. . .

Species	United States						Canada			Mexico		Central America
	SE	NE	C	SW	NW	A	E	C	W	N	S	
nebulosa (Walker)	X	X	X	X	...	X	...	X	...	...	...
pacifica (McDunnough)	X	...	...	...	...	...	...	...

NOTE: Berner (1975) transferred *bradleyi* from *Paraleptophlebia* to *Leptophlebia*. Figure 152 (*Paraleptophlebia* sp., Texas) illustrates gills of the *bradleyi* type; the nymphs from Texas are not *Paraleptophlebia* but *Leptophlebia*, possibly *L. bradleyi*. Since Berner's article was published after the production of this volume was under way, the transfer of *bradleyi* is not discussed in the text or represented in the keys.

Paraleptophlebia Lestage (figs. 29, 31, 149–152, 156, 161, 162, 290, 292–295, 300, 412)

Nymphal Characteristics. Body length: 6.5–10.0 mm. Body slender, slightly flattened. Head more or less hypognathous. In a few western species, mandibles with a greatly elongated and tusklike projection extending far anteriorly from labrum. Claws slender with a rather uniform row of fine denticles. Spinules present on apical margins of terga 1–10. Gills on abdominal segments 1–7 narrowly lanceolate, bifid. Illustration: Fig. 412.

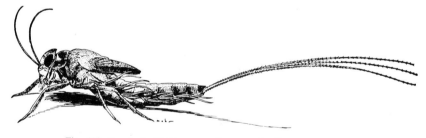

Fig. 412. *Paraleptophlebia praepedita*, nymph. (From Burks, 1953.)

Nymphal Habitat. Most common in shallow, fairly rapid streams of small to moderate size, but occurring also in larger rivers. The nymphs are often found in coarse gravel, but many species live among leaf debris where the current is slow to moderately swift. They also occur in riffles, on large logs or sticks, and in moss or vegetation. Other species occur in rapid portions of streams, hiding in crevices on the underside of rocks. In larger streams and rivers the nymphs are often found near banks on the exposed roots of terrestrial plants. As the nymphs mature, they migrate from the rapidly flowing water of the riffles to the quieter parts of the stream. The nymphs of two Oregon species are components of the insect stream drift.

Nymphal Habits. The nymphs are awkward swimmers, moving by undulatory movements. When taken out of water, they move in a snakelike fashion, wriggling from side to side. They appear to be herbivores, feeding on detritus and algae. Negatively phototropic.

Life History. The time required for nymphal development varies; in the northern part of the range species require a full year for development. In the extreme southern portion of the range less time is required. In Florida the nymphs may mature in six to eight months. When ready to transform, the mature nymph crawls a short distance above the water surface, and after a few moments the subimago appears. The subimaginal stage may last from twelve hours in the south to as much as forty-eight hours in the north. In the extreme southern portion of range emergence occurs throughout the year; in northern portion, however, emergence is restricted to the summer months, beginning as early as May and continuing into November. In two Oregon species, *P. temporalis* and *P. debilis*, there is a single generation each year. It appears that the time of hatching and the period of emergence are dependent on water flow and temperature. The eggs of *P. debilis* overwinter and hatch in the spring, with the adults emerging in the summer and fall. Adults are present in Oregon in March and as late as November. In Michigan and in the Rocky Mountains *P. debilis* is a member of the autumnal fauna, emerging in greatest numbers during September. *P. adoptiva* is one of the first mayflies to appear in Michigan in the spring, with the peak occurring during the first three weeks of May.

Adult Characteristics. Wing length: 5–12 mm. Markedly sexually dimorphic. Males usually with brownish head, thorax, and tip of abdomen; middle abdominal segments sometimes whitish, with or without obscure brownish markings, or sometimes dark brown. Females typically reddish or reddish to purplish brown, never with middle abdominal segments whitish. Wings clear, sometimes tinted with amber or dark brown. Hind wings broad and without a costal projection. Forelegs of male with tarsi 1½ times as long as tibiae. Male genitalia illustrated in figs. 290, 292–295. Penes diverse but never as in *Leptophlebia*. Three caudal filaments of equal length.

Mating Flights. Swarms have been observed in midafternoon on sunny days; they were composed mainly of males rising to a height of twenty feet and sometimes flying much lower at four to six feet. The swarms consist generally of ten to one hundred individuals. Other species have been observed swarming in the morning and at midday.

Generally mating flights occur over water, but Leonard and Leonard (1962) report mating flights in open areas some distance from the water. In Utah *P. memorialis* and *P. heteronea* form small swarms above bushes on hillsides as much as two hundred yards from a stream between midday and early afternoon. In Washington *P. vaciva* has been observed swarming in the sunlight at 10:00 A.M. and also in midafternoon. In Oregon *P. debilis* has been observed swarming as the sun set in late afternoon. As long as the sun was shining, only individual males or small groups of males were seen. Almost as soon as the sun ceased shining on the water surface, groups of thirty to forty males appeared and swarmed two to four feet above the water and close to shore. The males rose four or five feet and then dropped to within two feet of the water, occasionally falling onto the water surface. The females were seen about thirty minutes after the males began their flight; upon entering the swarm each female was seized by a male. Mating was completed during the time it took the pair to fall about three feet. Immediately after a pair separated, the female dropped to the water surface, released some eggs, then rose to fly a short distance to repeat the performance twice more; the spent female then flew to nearby vegetation. The males swarm only once. In Utah, as the days become colder in September and October, *P. debilis* swarms increasingly earlier in the day when the light intensity is much higher. In Washington and Oregon *P. bicornuta* swarms in late afternoon and early evening.

Taxonomy. Lestage, 1917. *Annales Biologie Lacustre* 8:340. Type species: *P. cincta* (Retzius); type locality: Central Europe. The identity of the species described by Retzius needs clarification. Generally, the adults of *Paraleptophlebia* should be identifiable to species from the existing keys; however, for many species it is necessary to compare the genitalia of the males with illustrations. For western species the papers by Day (1952, 1954a) should be consulted. The nymphs of this genus generally cannot be identified with any degree of accuracy. It is likely, as more rearings are completed, that a key to nymphs may be constructed, although regional keys may be the only feasible ones for many years. Berner (1975) has reviewed the southeastern species of *Paraleptophlebia*.

Distribution. Holarctic. In the Nearctic, distribution south to Florida, Texas, Arizona, New Mexico, and California.

Species	United States						Canada			Mexico		Central America
	SE	NE	C	SW	NW	A	E	C	W	N	S	
adoptiva (McDunnough)	X	X	X	...	...	...	X	...	...	...	...	...
altana Kilgore & Allen	...	...	...	X	...	...	...	...	...	...	...	...
assimilis (Banks)	X	...	...	...	...	...	...	...	...	...	...	...
associata (McDunnough)	...	...	...	...	X	...	...	...	...	...	...	...
bicornuta (McDunnough)	...	...	...	...	X	...	...	...	X	...	...	...

Species	United States						Canada			Mexico		Central America
	SE	NE	C	SW	NW	A	E	C	W	N	S	
bradleyi (Needham)	X	X	...	...	...	...	...	...	...	...	...	...
brunneipennis (McDunnough)	...	...	...	...	X	...	...	...	...	...	...	...
cachea Day	...	...	...	X	...	...	...	...	...	...	...	...
californica Traver	...	...	...	X	...	...	...	...	...	...	...	...
clara (McDunnough)	...	...	...	X	...	...	...	...	...	...	...	...
debilis (Walker)	X	X	X	X	X	...	X	X	X	...	...	...
falcula Traver	...	...	...	...	X	...	...	...	...	...	...	...
georgiana Traver	X	...	...	...	...	...	...	...	...	...	...	...
gregalis (Eaton)	...	...	...	X	X	...	...	...	X	...	...	...
guttata (McDunnough)	X	X	X	...	...	...	X	...	...	...	...	...
helena Day	...	...	...	X	...	...	...	...	...	...	...	...
heteronea (McDunnough)	...	...	...	X	X	...	...	...	X	...	...	...
jeanae Berner	X	...	...	...	...	...	...	...	...	...	...	...
memorialis (Eaton)	...	...	...	X	X	...	...	...	X	...	...	...
moerens (McDunnough)	X	X	X	...	...	...	X	...	...	...	...	...
mollis (Eaton)	X	X	X	...	...	...	X	...	...	...	...	...
ontario (McDunnough)	...	X	X	...	...	...	X	...	...	...	...	...
packi (Needham)	...	...	...	X	...	...	...	...	...	...	...	...
placeri Mayo	...	...	...	X	...	...	...	...	...	...	...	...
praepedita (Eaton)	...	X	X	...	...	...	X	...	...	...	...	...
quisquilia Day	...	...	...	X	...	...	...	...	...	...	...	...
rufivenosa (Eaton)	...	...	...	...	X	...	...	...	X	...	...	...
sticta Burks	...	...	X	...	...	...	...	...	...	...	...	...
strigula (McDunnough)	...	X	X	...	...	...	X	...	...	...	...	...
swannanoa (Traver)	X	...	...	...	...	...	...	...	...	...	...	...
temporalis (McDunnough)	...	...	...	...	X	...	...	...	X	...	...	...
vaciva (Eaton)	...	...	...	...	X	...	...	...	...	...	...	...
volitans (McDunnough)	X	X	...	...	...	...	X	...	...	...	...	...
zayante Day	...	...	...	X	...	...	...	...	...	...	...	...
Species?	...	...	...	...	...	X	...	...	...	...	...	...

NOTE: Berner (1975) transferred *bradleyi* from *Paraleptophlebia* to *Leptophlebia*. Figure 152 (*Paraleptophlebia* sp., Texas) illustrates gills of the *bradleyi* type; the nymphs from Texas are not *Paraleptophlebia* but *Leptophlebia*, possibly *L. bradleyi*. Since Berner's article was published after the production of this volume was under way, the transfer of *bradleyi* is not discussed in the text or represented in the keys.

Thraulodes Ulmer (figs. 158, 159, 301, 305, 413)

Nymphal Characteristics. Body length: 5–12 mm. Body flattened. Head flattened, squarish in dorsal aspect, about as wide as long; prognathous. Labrum much narrower than head. Mandibles with a row of hairs from near middle of outer margin to base of outer incisors. Seven pairs of double lanceolate gills, diminishing in size posteriorly. Posterolateral angles of abdominal segments 2–9 produced as slender spines, but short on segments 2–3 or 2–4. Illustration: Fig. 413.

Nymphal Habitat. In Central America the nymphs are found commonly wherever well-aerated water flows over rocky stream bottoms; some rivers that are milky in color are highly productive. The nymphs have also been taken in Honduran streams with clear water, sandy bottoms, and few rocks. In the upper parts of the Amazon drainage in Peru, the nymphs are found on the stems of emergent plants in large rivers. In Arizona the nymphs occur in small or medium-sized clear streams. The nymphs are most abundant in water that is three to six inches deep, but they occupy a wide range of habitats.

Nymphal Habits. The nymphs are rather poor swimmers and creep over the surface as do the nymphs of *Paraleptophlebia* or *Choroterpes*, although in shape the *Thraulodes* nymphs are much more flattened.

Life History. Not described.

Adult Characteristics. Wing length: 5.5–14.0 mm. Eyes of male turbinate, may appear circular or oval from dorsal aspect; turbinate portion set on short stout stalk; usually contiguous dorsally. Forelegs of male more than three-fourths length of body; femora usually with distinct subapical dark bands; fore tarsi shorter than tibiae; basal tarsal segment shorter than second segment. Costal brace of forewings usually darkened, especially near apex. Costal projection of hind wings usually blunt, rarely acute, usually at about midpoint of length; vein Sc continues short distance beyond angulation; MP of hind wings forked. Male genitalia illustrated in fig. 305. Penes each with a spear-like process arising dorsally from distal margin, the points of the spears directed inward toward each other. Caudal filaments often with alternating dark and light bands.

Mating Flights. The adults swarm at dusk or after dark. The male imagoes often gather around lights in considerable numbers.

Taxonomy. Ulmer, 1920b. *Archiv für Naturgeschichte* 85:33. Type species: *T. laetus* (Eaton); type locality: New Granada, Colombia. The male adults of *Thraulodes* should be identifiable to species using the recent monograph of the genus by Traver and Edmunds (1967). Koss (1966) has prepared a simpler key to the United States species of the genus. The nymphs are poorly known, and no key is available. In Mexico and Central America nymphs of this genus are among the most widespread and most commonly encountered mayflies,

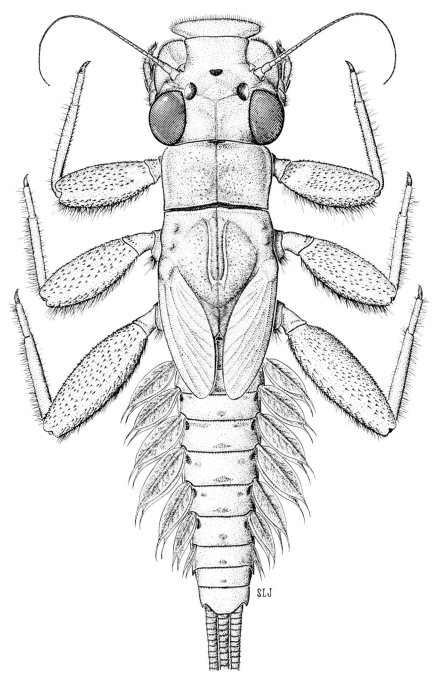

Fig. 413. *Thraulodes* sp., nymph. (From Edmunds, Allen, and, Peters, 1963.)

and it is probable that a number of species remain to be described in these areas.

Distribution. Neotropical and Nearctic. In the Nearctic, extending north to Arizona, New Mexico, and Texas.

Species	United States						Canada			Mexico		Central America
	SE	NE	C	SW	NW	A	E	C	W	N	S	
arizonicus McDunnough	...	...	...	X	...	...	...	...	...	...	...	...
brunneus Koss	...	...	...	X	...	...	...	...	...	...	...	...
centralis Traver	...	...	...	...	...	...	...	...	...	...	...	X
ephippiatus Traver & Edmunds	...	...	...	...	...	...	...	...	...	X		...
gonzalesi Traver & Edmunds	...	...	X	...	...	...	...	...	...	X	...	...
hilaris (Eaton)	...	...	...	...	...	...	...	...	...	...	...	X
hilaroides Traver	...	...	...	...	...	...	...	...	...	...	...	X
humeralis Navas	...	...	...	...	...	...	...	...	...	X		...
irretitus Navas	...	...	...	...	...	...	...	...	...	...	...	X
lepidus (Eaton)	...	...	...	...	...	...	...	...	...	...	...	X
lunatus Traver & Edmunds	...	...	...	...	...	...	...	...	...	X	...	...
mexicanus (Eaton)	...	...	...	...	...	...	...	...	...	X		...
packeri Traver & Edmunds	...	...	...	...	...	...	...	...	...	...	...	X
prolongatus Traver	...	...	...	...	...	...	...	...	...	...	...	X
salinus Kilgore & Allen	...	...	...	X	...	...	...	...	...	...	...	...
spangleri Traver & Edmunds	...	...	...	...	...	...	...	...	...	...	X	X
speciosus Traver	...	...	...	X	...	...	...	...	...	...	...	...
valens (Eaton)	...	...	...	...	...	...	...	...	...	...	...	X
zonalis Traver & Edmunds	...	...	...	...	...	...	...	...	...	...	...	X

Thraulus Eaton

Various species of wide geographic occurrence have been assigned to this genus, but most American species assigned here earlier have been transferred to other genera, principally to *Thraulodes*, *Traverella*, and *Homothraulus*. Demoulin (1955) has suggested that all the American species remaining in *Thraulus* should be transferred to *Homothraulus*, but we agree with Traver (1960b), with whom Demoulin now concurs, that not all of these belong in *Homothraulus*. Therefore, two Central American species, *roundsi* Traver and *sallei* Navas, remain in the genus. Although it is certain that neither species represents the real genus *Thraulus*, the correct generic assignment requires further knowledge of the Leptophlebiidae, and we do not attempt to deal further with these species herein. The true genus *Thraulus* is African, circum-Mediterranean, and Oriental, extending to New Guinea.

Traverella Edmunds (figs. 135, 139, 309, 333)

Nymphal Characteristics. Body length: 7–10 mm. Body flattened. Head rectangular, prognathous; eyes dorsal; labrum as wide as head; a distinct projection mesally on clypeus at base of labrum; maxillary palpi lie along side of head visible dorsally. Claws with about twelve denticles, those on forelegs short and wartlike, those of middle and hind legs usually well developed. Abdominal segments 8 and 9 with posterolateral projections. Gills on abdominal segments 1–7 double; gills on segment 1 largest, others diminishing in size to gill 7; bilamellate, each lamella of gills 1–5 with a fringed margin, posterior lamellae of each pair about two-thirds to three-fourths as large as anterior lamellae. Illustration: Edmunds, 1948b, fig. 16.

Nymphal Habitat. In rapids of medium-sized to large rivers where the nymphs may be found on the underside of rocks. The mature nymphs may occur on certain rocks in large numbers (more than a hundred per square foot of surface) while most nearby rocks in a similar habitat may have only one or two nymphs per square foot.

Nymphal Habits. In behavior the nymphs are similar to nymphs of the Heptageniidae. They move freely forward, backward, or to the side. They are poor swimmers, moving about primarily by crawling. They feed by filtering algae and detritus from the current.

Life History. The adults emerge in Utah during late August and probably throughout September. The subimagoes emerge at about 7:10 P.M. when the light intensity is reduced to approximately 5 footcandles. During the next twenty minutes many individuals emerge, but emergence stops abruptly when the light intensity drops to 1 footcandle. In the United States emergence occurs primarily in August and September. In Mexico and Central America emergence takes place over a longer season.

Adult Characteristics. Wing length: 6–10 mm. Eyes of male large, nearly contiguous dorsally. Forelegs of male about two-thirds body length; tarsi slightly shorter than tibiae. Hind wings with an acute costal projection at about midpoint of length. Male genitalia illustrated in fig. 309; a pair of rodlike projections dorsal to forceps.

Mating Flights. In Utah the adults swarm over the shores of rivers as soon as the sun shines on the area in the morning. The flight continues until nearly midday, depending on the temperature. Swarming occurs from ten to twenty feet above the ground, especially near vegetation along river banks. The wings beat on both the rising and descending parts of the flight.

Taxonomy. Edmunds, 1948b. *Proc. Biol. Soc. Wash.* 61:141–42. Type species: *T. albertana* (McDunnough); type locality: Medicine Hat, Alberta. Only two species in the United States are described as adults and four as nymphs. In Mexico and Central America two species are named from adults, and three nymphal forms are designated by letters. A large series of adults of *T. castanea* are known to us but not described. Allen (1973) has compiled a key to the nymphs.

Distribution. Neotropical and Nearctic. In the Nearctic the genus extends up warm rivers north to Washington, Alberta, Saskatchewan, and Ohio.

Species	United States						Canada			Mexico		Central America
	SE	NE	C	SW	NW	A	E	C	W	N	S	
albertana (McDunnough)	...	...	X	X	X	...	...	X	X	...	...	...
castanea Kilgore & Allen	...	...	...	X	...	...	...	...	...	...	...	...
lewisi Allen	...	...	X	...	...	...	...	...	...	...	...	...
presidiana (Traver)	...	...	X	...	...	...	...	...	...	X	...	...
primana (Eaton)	...	...	...	...	...	...	...	...	...	...	X	X
versicolor (Eaton)	...	...	...	...	...	...	...	...	...	...	...	X

Ulmeritus Traver (figs. 141, 142, 291, 302)

Nymphal Characteristics. Body length: 6.5–9.0 mm. Body somewhat flattened. Head prognathous. Labrum less than half width of head. Claws with denticles progressively larger toward apex. Gill lamellae double on segments 1–7, margins fringed; in some species fringes on all edges of lamellae, in others fringes absent on lateral margins of lamellae. Posterolateral projections rather large on posterior segments. Illustrations: Parts in Traver, 1956b.

Nymphal Habitat. Nymphs occur principally in rivers and streams, but one collection in Panama was obtained from a pool.

Nymphal Habits. Unknown.

Life History. Unknown.

Adult Characteristics. Wing length: 6.5–9.0 mm. Eyes of male contiguous dorsally. Costal projection of hind wings poorly developed, rounded, about one-third distance from base of wing. Vein MP of hind wings forked. Genitalia of male as in fig. 291.

Mating Flights. Peters (personal communication) reports that Brazilian *Ulmeritis* fly after dark. The males fly swiftly and are agile, often avoiding attempts to catch them with a net.

Taxonomy. Traver, 1956b. *Proc. Entomol. Soc. Wash.* 58:2. Type species: *U. carbonelli* Traver; type locality: Cuareim River, Artigas Province, Sepultras, Uruguay. The Central American species are known only from nymphs from Panama and Costa Rica. The Panama nymphs have gills similar to those of *U. carbonelli* in that they are fringed on all margins. The nymphs from Costa Rica have no fringes on the lateral margins of the lamellae and may represent a different genus or subgenus. None of the Central American nymphs is assignable to species.

Family EPHEMERELLIDAE

Subfamily Ephemerellinae

The Ephemerellidae probably arose from the phyletic line leading to the Leptophlebiidae and the Ephemeroidea. The most closely related extant mayflies are the Tricorythidae, which appear to be derived from Ephemerellidae. The Ephemerellidae and the Tricorythidae of North and Central America do not indicate this relationship as well as do the African genera of the two families. The family Ephemerellidae is assigned to the Leptophlebioidea, but because the Leptophlebiidae appear to have given rise to the Ephemeroidea, some workers prefer to isolate the Ephemerellidae and the Tricorythidae in a separate superfamily, Ephemerelloidea.

The family is widespread on the continents, but it is absent from New Zealand. There are three subfamilies. The enigmatic Melanemerellinae are known only from Brazil; the Teloganodinae are primarily Ethiopian and Oriental but extend also to Queensland in Australia, to southern China in the Palearctic, and to Madagascar. The Ephemerellinae are abundant and diverse in the Holarctic but extend into the Oriental region. Furthermore, there is a highly questionable record of *Ephemerella* from Madagascar.

The genus *Ephemerella* is the only genus of the family in the Nearctic. See *Ephemerella* for characteristics and distribution in North America.

Ephemerella Walsh

Nymphal Characteristics. Body length: 5–15 mm. Variable in body form; some forms slender and streamlined, others flattened ventrally with dorsum arched and with broad, flattened femora. Many have dorsal tubercles on head, thorax, and/or abdomen. Gills rudimentary on segment 1 or absent from segment 1, absent from segment 2 or segments 2–3; bilamellate on segments 3–7 or 4–7; ventral lamellae subdivided. If present only on segments 4–7, gills on segment 4 operculate or semioperculate. Lateral extensions usually present on the abdominal segments, bearing posterolateral spines of variable development. Caudal filaments with spines, setae, or both. Nymphs of some species highly variable in color. Illustrations: Figs. 414–417.

Nymphal Habitat. The nymphs occur in a wide variety of habitats — in lakes where there is wave action along the shore, in slow-flowing to virtually stagnant streams, and in swift streams with rapids. Wherever they occur, they usually seek protection in crevices of rocks or in vegetation where the flow is reduced. In two western forms the nymphs are adapted to life in torrential conditions; the abdomen is modified to form a sucker-disc that permits the nymph to cling to rocks in the swiftest water. Forms occurring in quiet water are very often found at the stream banks, where they bury themselves close to the bases of plants and among roots in shallow water. Some nymphs live deep within growths of moss.

Nymphal Habits. Most nymphs cling very tightly to the objects to which they are attached. When removed from the water, they remain quiescent; as they dry out, some slow, deliberate movements can be seen. Those living in moss will leave the moss as it begins to dry. In some forms the nymphs flick the tail forward over the abdomen, then straighten it, repeating the act continually. In others there is little or no movement, and they are difficult to detect. The nymphs are awkward swimmers, using undulatory movements of the abdomen for locomotion; they almost immediately seek an object to which they can cling, and once attached they blend with the background. Those nymphs with operculate gills circulate water under the gills by raising one or both covers and vibrating the other pairs of gills. Omnivorous.

Life History. The time required for development for most species is nearly one year. Most western species are known to develop in nine to eleven months. In Florida emergence takes place in late winter, spring, and early summer. Most western species emerge from May to September. In some species emergence occurs as the nymph floats at the surface of the water, and almost immediately the subimago flies away. In others the nymph may crawl partially out of the water onto some attached object and transform. Emergence normally takes place in late afternoon, sometimes occurring after dark. The subimago stage may last from twenty-two to thirty hours.

Adult Characteristics. Wing length: 6–19 mm. Compound eyes of male almost contiguous dorsally. Forelegs of male with tarsal segments 2 and 3 variable, segment 4 about three-fourths the length of segment 3; apical segment generally less than half the length of segment 4. Hind tarsi with four clearly differentiated segments. Claws dissimilar on all tarsi. Forewings with costal crossveins absent or with only a few weakly developed apical ones; stigmatic crossveins weak and slightly anastomosed; other crossveins weak;

short basally detached marginal intercalaries present between veins along entire outer margin of wings. Hind wings with slightly rounded costal projection; crossveins weak. Male genitalia illustrated in figs. 311–316, 318–319, 321–325, 378; a single short apical segment on genital forceps. Three caudal filaments.

Mating Flights. The flights generally occur at dusk, and few reports of observations of them have been published. The swarms of some species are small, and the adults fly above or near streams. Many of the species may fly at high altitudes. Observations have been made of a few species flying at six to twenty feet above the ground. In the west *E. tibialis* have been observed flying fifty feet or more above the ground on occasions and as low as three feet above the roof of a car; *E. coloradensis* may fly in huge swarms nearly a hundred feet in the air. When ready to oviposit, the female extrudes the eggs in a compact spherical mass attached to the underside of her abdomen, with the abdomen partially reflexed over the mass. The female flies back and forth over the stream surface, finally dropping to the water and releasing the eggs.

Taxonomy. Walsh, 1862. *Proc. Acad. Natur. Sci. Philadelphia* 13–14:377. Type species: *E. excrucians* Walsh; type locality: Rock Island, Illinois. This is the only primarily North American genus of any size for which a reasonably recent generic revision is available. The subgenera were characterized and keyed by Edmunds (1959a), but the somewhat longer and more utilitarian key by Allen and Edmunds (1965) is improved and incorporated in this book. See the various subgenera for reference to the revision of each subgenus. Additional species have been described by Jensen and Edmunds (1966), Allen (1968), and Allen and Collins (1968).

Distribution. Holarctic and Oriental, with a dubious record from Madagascar. In the Nearctic, very abundant and widespread in the United States and Canada south to Baja California.

Subgenus *Attenella* Edmunds (figs. 311, 414)

Nymphal Characteristics. Body length: 5.0–9.5 mm. Claws with denticles. Abdomen with moderately well developed paired submedian tubercles on terga 2–9, 3–9, or 3–8; gills present on segments 4–7 with rudimentary gills on segment 1. Abdominal segments 8 and 9 subequal in length. Illustration: Fig. 414.

Adult Characteristics. Wing length: 6 mm. Terminal segment of forceps of male six times as long as broad. Male genitalia as in fig. 311. Unassociated females cannot be identified to subgenus.

Taxonomy. Edmunds, 1959c. *Ann. Entomol. Soc. Amer.* 52:546. Type Species: *E. (A.) attenuata* McDunnough; type locality: Ottawa Golf Club (near Hull), Quebec. Allen and Edmunds (1961b) key and describe the known

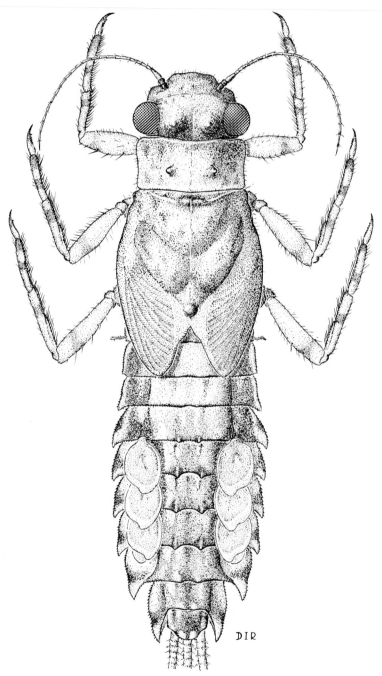

Fig. 414. *Ephemerella* (*Attenella*) *attenuata*, nymph. (From Allen and Edmunds, 1961b.)

adults and nymphs of this subgenus. Originally the subgenus was named *Attenuatella*, but since this name was preoccupied, the subgenus was renamed *Attenella*.

Distribution. Nearctic. Species distributed in both eastern and western North America. In eastern North America, known from Nova Scotia, New Brunswick, and Quebec south to northern Florida; in western North America, known from British Columbia and Alberta south to California and New Mexico. *E. (A.) margarita* has populations in both eastern and western North America.

	United States						Canada			Mexico		Central
Species	SE	NE	C	SW	NW	A	E	C	W	N	S	America
attenuata												
McDunnough	X	X	X	...	...	...	X	...	...	...	...	...
delantala												
Mayo	...	...	...	X	X	...	...	...	...	...	...	...
margarita												
Needham	...	X	...	X	X	...	...	...	X	...	...	...
soquele Day..........	...	...	...	X	X	...	...	...	...	...	...	...

Subgenus *Caudatella* Edmunds (fig. 378)

Nymphal Characteristics. Body length: 5–11 mm. Abdomen with moderately developed to well-developed paired submedian tubercles on terga 1–10, 1–9, or 2–9. Gills present on abdominal segments 3–7. Cerci one-fourth to three-fourths length of terminal filament. Illustrations: Allen and Edmunds, 1961a, fig. 1.

Adult Characteristics. Wing length: 6.5–12.0 mm. Cerci one-fourth to three-fourths length of terminal filament. Male genitalia as in fig. 378.

Taxonomy. Edmunds, 1959c. *Ann. Entomol. Soc. Amer.* 52:546. Type species: *E. (C.) heterocaudata* McDunnough; type locality: Yellowstone National Park, Wyoming. Allen and Edmunds (1961b) key and describe the species and the subspecies for both adults and nymphs.

Distribution. Nearctic. Restricted to northwestern North America from British Columbia and Alberta south to central California, northern Idaho, and northwestern Wyoming. Most of the species have narrow environmental tolerances and are restricted to streams scattered in the Cascades, the Sierra Nevada, the Coast Range, and the Rocky Mountains.

	United States						Canada			Mexico		Central
Species	SE	NE	C	SW	NW	A	E	C	W	N	S	America
cascadia Allen												
& Edmunds	...	...	...	...	X	...	...	...	...	...	...	...
edmundsi Allen	...	...	...	...	X	...	...	...	...	...	...	...

Species	United States						Canada			Mexico		Central America
	SE	NE	C	SW	NW	A	E	C	W	N	S	
heterocaudata californica Allen & Edmunds ...	...	...	...	...	X	...	...	...	...	...	...	...
heterocaudata circia Allen & Edmunds ...	...	...	...	...	X	...	...	...	...	...	...	...
heterocaudata heterocaudata McDunnough	...	...	...	X	X	...	...	...	X	...	...	...
hystrix Traver	...	...	...	X	X	...	...	...	X	...	...	...
jacobi McDunnough	...	...	...	...	X	...	...	...	X	...	...	...

Subgenus *Dannella* Edmunds (figs. 183, 324)

Nymphal Characteristics. Body length: 6–8 mm. Claws without denticles. Abdomen without dorsal paired submedian tubercles; gills present on segments 4–7 with rudimentary gills present on segment 1; gills on segment 4 semioperculate. Illustrations: Allen and Edmunds, 1962a, fig. 1.

Adult Characteristics. Wing length: 6.0–6.5 mm. Fore tarsi with segment 3 longer than segment 2. Terminal segment of forceps of male about as broad as long, second segment slender; penes expanded apically and without subapical projections or dorsal or ventral spines. Male genitalia as in fig. 324. Unassociated females cannot be identified to subgenus.

Taxonomy. Edmunds, 1959c. *Ann. Entomol. Soc. Amer.* 52:546. Type species: *E. (D.) simplex* McDunnough; type locality: Laprairie, Quebec. The publication of Allen and Edmunds (1962a) provides a key to the nymphs of the two species in this subgenus. R. K. Allen (personal communication) has informed us very recently that he has seen an undescribed nymph of a species of the subgenus *Dannella* with paired tubercles on the abdomen. This species, from Lake Huron, does not key out to *Dannella* at couplet 77. It could possibly key to *Attenella*, but in general facies it resembles the nymph of *E. (Dannella) simplex*, with segments 7 and 8 somewhat shortened and the margins of the abdomen having numerous setae.

Distribution. Nearctic. Restricted to eastern North America from Manitoba and New Brunswick south to Illinois and Florida. *E. simplex* is more common and widespread than *E. lita* Burks. The latter species is confined to the United States.

Subgenus *Drunella* Needham (figs. 52, 167, 171, 312, 313)

Nymphal Characteristics. Body length: 6–15 mm. Head, thorax, and abdomen with or without well-developed tubercles; when absent, the ventral (leading) edge of the fore femora has distinct tubercles, or abdominal sterna 3–8 have an adhesive disc of long hair. Gills present on abdominal segments 3–7. Illustration: Allen and Edmunds, 1962b, fig. 1.

Adult Characteristics. Wing length: 6–19 mm. Terminal segment of genital forceps of male twice as long as broad, and long second segment distinctly incurved or strongly bowed. Male genitalia as in figs. 312, 313. Unassociated females cannot be identified to subgenus.

Taxonomy. Needham, 1905. *New York State Mus. Bull.* 84:42. Type species: *E. (D.) grandis* Eaton; type locality: Colorado. Allen and Edmunds (1962b) provide keys and descriptions to the North American species in this subgenus.

Distribution. Holarctic. In the Nearctic, widely distributed in western North America from Alaska, British Columbia, and Alberta south to Baja California and New Mexico; in eastern North America, known from Ontario, Quebec, and Nova Scotia south to Michigan, South Carolina, and Georgia.

Species	United States						Canada			Mexico		Central America
	SE	NE	C	SW	NW	A	E	C	W	N	S	
allegheniensis Traver	X	X	...	...	...	...	...	...	...	...	...	...
coloradensis Dodds	...	...	...	X	X	X	...	...	X	...	...	...
conestee Traver	X	...	...	...	...	...	...	...	...	...	...	...
cornuta Morgan	X	X	X	...	...	...	X	...	...	...	...	...
cornutella McDunnough	X	X	X	...	...	...	X	...	...	...	...	...
doddsi Needham	...	...	...	X	X	X	...	...	X	...	...	...
flavilinea McDunnough	...	...	...	X	X	...	...	...	X	X	...	...
grandis flavitincta McDunnough	...	...	...	X	X	...	...	...	X	...	...	...
grandis grandis Eaton	...	...	...	X	X	...	...	...	...	...	...	...
grandis ingens McDunnough	...	...	...	X	X	...	...	...	X	...	...	...
lata Morgan	X	X	X	...	...	...	X	...	...	...	...	...
longicornis Traver	X	X	...	...	...	...	...	...	...	...	...	...
pelosa Mayo	...	...	...	X	X	...	...	...	...	...	...	...
spinifera Needham	...	...	...	X	X	...	...	...	X	...	...	...
tuberculata Morgan	X	X	...	...	...	...	X	...	...	...	...	...
walkeri Eaton	X	X	X	...	...	...	X	...	...	...	...	...
wayah Traver	X	...	...	...	...	...	...	...	...	...	...	...

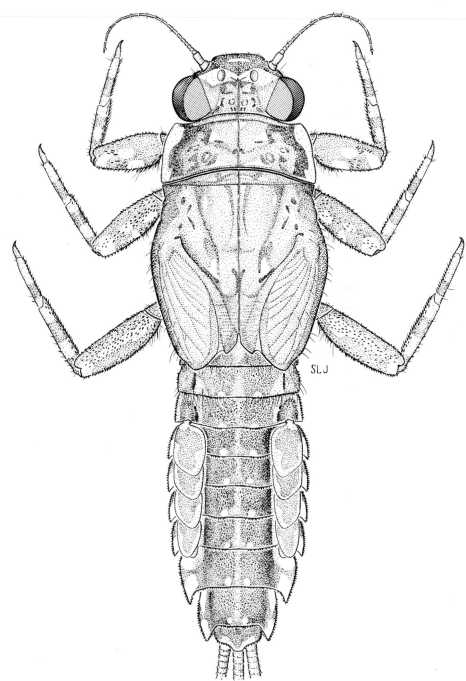

Fig. 415. *Ephemerella (Ephemerella) inermis*, nymph. (From Allen and Edmunds, 1965.)

Subgenus *Ephemerella* Walsh, sensu stricto (figs. 170, 173, 174, 225, 314–316, 320, 322, 415)

Nymphal Characteristics. Body length: 6–14 mm. Maxillary palpi well developed. Abdomen with or without paired dorsal submedian tubercles; gills present on segments 3–7. Caudal filaments with or without spines at apex of each segment and with heavy intersegmental setae. Illustration: Fig. 415.

Adult Characteristics. Wing length: 6–13 mm. Penes of male usually with dorsal or ventral spines; when spines are absent, either the penes have long lateral apical lobes and vestiges of the nymphal gills are absent, or abdominal sterna 2–7 have reddish brown to dark brown medially notched rectangular markings. Male genitalia as in figs. 314–316, 322. Unassociated females cannot be identified to subgenus.

Taxonomy. Walsh, 1862. *Proc. Acad. Natur. Sci. Philadelphia* 13–14:377. Type species: *E. (E.) excrucians* Walsh; type locality: Rock Island, Illinois. The keys and descriptions of Allen and Edmunds (1965) permit identification of the North American species in this subgenus. A subsequent paper by Allen (1968) keys the western North American species of the subgenus, including several species described since 1965.

Distribution. Holarctic. Widespread in the Nearctic from Alaska, Alberta, Manitoba, Ontario, Quebec, and Newfoundland south to California, New Mexico, Oklahoma, Alabama, and Florida.

Species	United States						Canada			Mexico		Central America
	SE	NE	C	SW	NW	A	E	C	W	N	S	
alleni Jensen & Edmunds	...	...	...	...	X	...	...	...	...	...	...	...
altana Allen	...	...	...	X	...	...	...	...	...	...	...	...
argo Burks	...	...	X	...	...	...	...	...	...	...	...	...
aurivillii Bengtsson	...	X	X	X	X	X	X	...	X	...	...	...
berneri Allen & Edmunds	X	...	...	...	...	...	...	...	...	...	...	...
catawba Traver	X	...	...	...	...	...	...	...	...	...	...	...
choctawhatchee Berner	X	...	...	...	...	...	...	...	...	...	...	...
crenula Allen & Edmunds	X	...	...	...	...	...	...	...	...	...	...	...
dorothea Needham	X	X	X	...	...	...	X	...	...	...	...	...
excrucians Walsh.............	...	X	X	...	...	...	X	...	...	...	...	...
fratercula McDunnough	X?	...	...	...	...	...	X	...	...	...	...	...
hispida Allen & Edmunds	X	...	...	...	...	...	...	...	...	...	...	...

Species	United States SE	NE	C	SW	NW	Canada A	E	C	W	Mexico N	S	Central America
inconstans Traver	X			X								
inermis Eaton				X	X	X			X			
infrequens McDunnough				X	X				X			
invaria (Walker)	X	X	X		X				X			
lacustris Allen & Edmunds				X								
maculata Traver				X								
mollitia Seemann			X									
needhami McDunnough		X	X						X			
ora Burks			X									
rama Allen			X									
rossi Allen & Edmunds			X						X			
rotunda Morgan	X	X	X						X			
septentrionalis McDunnough	X	X										
simila Allen & Edmunds	X											
subvaria McDunnough		X	X						X			
verruca Allen & Edmunds					X							

Subgenus *Eurylophella* Tiensuu (figs. 177, 325, 416)

Nymphal Characteristics. Body length: 6–11 mm. Head usually with paired occipital tubercles. Abdomen with paired dorsal submedian tubercles present on segments 1–9 or 1–10; operculate gills present on segments 4–7, with rudimentary gill present on segment 1. Illustration: Fig. 416.

Adult Characteristics. Wing length: 6–11 mm. Fore tarsi with segment 3 shorter than segment 2. Terminal segment of forceps of male about as broad as long, second segment thick; penes expanded basally, narrow apically, and without subapical projections or dorsal or ventral spines. Male genitalia as in fig. 325. Unassociated females cannot be determined to subgenus.

Taxonomy. Tiensuu, 1935. *Ann. Entomol. Fenn.* 1:20. Type species: *E.* (*E.*) *brunnescens* (Tiensuu); type locality: Kurkijoki, Lake Ladoga (Finland), Karelian A.S.S.R. Allen and Edmunds (1963b) provide keys and descriptions of the North American and European species in this subgenus.

Distribution. Holarctic. In the Nearctic, distribution in western North America from British Columbia south to California and in eastern North

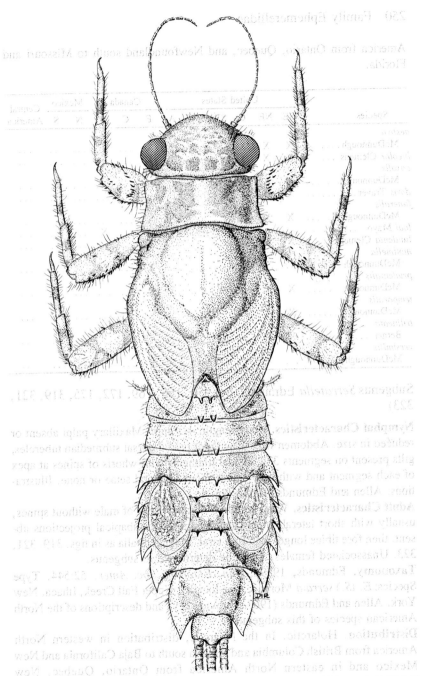

Fig. 416. *Ephemerella* (*Eurylophella*) *bicolor*, nymph. (From Allen and Edmunds, 1963b.)

America from Ontario, Quebec, and Newfoundland south to Missouri and Florida.

Species	United States						Canada			Mexico		Central America
	SE	NE	C	SW	NW	A	E	C	W	N	S	
aestiva												
McDunnough	X	X	X	...	...	...	X	...	...	...	...	...
bicolor Clemens	X	X	X	...	...	...	X	...	...	...	...	...
coxalis												
McDunnough	X	X	X	...	...	...	X	...	...	...	...	...
doris Traver	X	...	...	...	...	...	...	...	...	...	...	...
funeralis												
McDunnough	X	X	X	...	...	...	X	...	...	...	...	...
lodi Mayo	...	...	...	X	X	...	...	...	X	...	...	...
lutulenta Clemens	X	X	X	...	...	...	X	...	...	...	...	...
minimella												
McDunnough	X?	...	X	...	...	...	X	...	...	...	...	...
prudentalis												
McDunnough	X	X	...	...	...	...	X	...	...	...	...	...
temporalis												
McDunnough	X	X	X	...	...	...	X	X	...	...	...	...
trilineata												
Berner	X	...	...	...	...	...	...	...	...	...	...	...
versimilis												
McDunnough	...	X	X	...	...	...	X	...	...	...	...	...

Subgenus *Serratella* Edmunds (figs. 50, 168, 169, 172, 175, 319, 321, 323)

Nymphal Characteristics. Body length: 4–9 mm. Maxillary palpi absent or reduced in size. Abdomen with or without paired dorsal submedian tubercles; gills present on segments 3–7; caudal filaments with whorls of spines at apex of each segment and with only sparse intersegmental setae or none. Illustrations: Allen and Edmunds, 1963a, fig. 1.

Adult Characteristics. Wing length: 5–9 mm. Penes of male without spines, usually with short lateral subapical projections; if subapical projections absent, then fore tibiae longer than fore tarsi. Male genitalia as in figs. 319, 321, 323. Unassociated females cannot be determined to subgenus.

Taxonomy. Edmunds, 1959a. *Ann. Entomol. Soc. Amer.* 52:544. Type Species: *E. (S.) serrata* Morgan; type locality: Upper Fall Creek, Ithaca, New York. Allen and Edmunds (1963a) provide keys and descriptions of the North American species of this subgenus.

Distribution. Holarctic. In the Nearctic, distribution in western North America from British Columbia and Alberta south to Baja California and New Mexico and in eastern North America from Ontario, Quebec, New Brunswick, and Nova Scotia south to Florida, Mississippi, and Arkansas.

Species	United States						Canada			Mexico		Central America
	SE	NE	C	SW	NW	A	E	C	W	N	S	
carolina Berner & Allen	X	...	...	...	...	...	...	...	...	·>·	...	...
deficiens Morgan	X	X	X	...	...	...	X	...	...	...	...	...
frisoni McDunnough	X	...	X	...	...	...	...	...	...	...	...	...
levis Day	...	...	...	X	...	...	...	...	...	...	...	...
micheneri Traver	...	...	...	X	X	...	...	...	...	X	...	...
molita McDunnough	...	...	...	...	...	...	X	...	...	...	...	...
sequoia Allen & Collins	...	...	...	X	...	...	...	...	...	...	...	...
serrata Morgan	X	X	...	...	...	...	X	...	...	...	...	...
serratoides McDunnough	X	X	...	...	...	...	X	...	...	...	...	...
sordida McDunnough	X	X	X	...	...	...	X	...	...	...	...	...
spiculosa Berner & Allen	X	...	...	...	...	...	...	...	...	...	...	...
teresa Traver	...	...	...	X	X	...	...	...	...	...	...	...
tibialis McDunnough	...	...	...	X	X	X	...	...	X	...	...	...
velmae Allen & Edmunds	...	...	...	...	X	...	...	...	...	...	...	...

Subgenus *Timpanoga* Needham (figs. 176, 317, 318, 417)

Nymphal Characteristics. Body length: 12–15 mm. Body strongly depressed. Head with broad frontal shelf. Claws without denticles. Abdominal segments with long posterolateral projections; gills present on segments 4–7, gills on segment 4 operculate. Illustration: Fig. 417.

Adult Characteristics. Wing length: 12–14 mm. Abdomen with vestiges of nymphal gills on segments 4–7; well-developed posterolateral projections on segments 8 and 9. Male genitalia as in figs. 317, 318.

Taxonomy. Needham, 1927. *Ann. Entomol. Soc. Amer.* 20:108. Type species: *E. (T.) hecuba* (Eaton); type locality: Manitou, Colorado. This subgenus contains a single species with two subspecies. Allen and Edmunds (1959) provide characters to distinguish the nymphs of the subspecies; *E. h. pacifica* Allen and Edmunds is distributed west of the Cascades and the Sierra Nevada, and *E. h. hecuba* is distributed east of the Cascades and the Sierra Nevada through the Rocky Mountains.

Distribution. Nearctic. Northwestern North America from British Columbia and Alberta south to California and New Mexico. The nymphs occur in large, moderately swift trout streams.

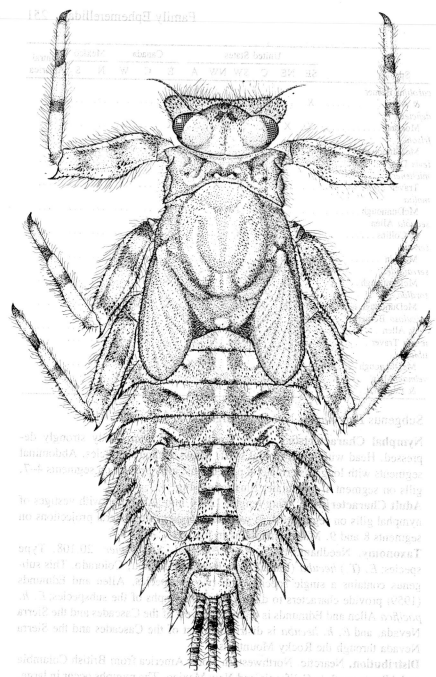

Fig. 417. *Ephemerella (Timpanoga) h. hecuba*, nymph. (From Allen and Edmunds, 1959.)

Family TRICORYTHIDAE

This family has many specialized characters and appears to have arisen from proto-teloganodine Ephemerellidae; it is usually assigned with the Ephemerellidae to the Leptophlebioidea. The family is extremely diverse in Africa, where all the subfamilies (Tricorythinae, Ephemerythinae, Machadorhythinae, Dicercomyzinae, and Leptohyphinae) are present. The Tricorythinae extend through the Oriental region. Leptohyphinae is the only subfamily occurring in the Americas.

Subfamily Leptohyphinae

This subfamily reaches its maximum diversity in South America. The two North American genera and the one Central American genus are northward extensions of the Neotropical fauna in which *Leptohyphes* and *Tricorythodes* are each represented by many species; one additional species of *Haplohyphes* occurs in Peru.

Nymphal Characteristics. Body robust, with large mesothorax and relatively short abdomen. Gills on abdominal segments 2–6; gills on segment 2 operculate to semioperculate, oval, or triangular, covering all or most of gills 3–6. Three short caudal filaments, robust at base, with rows of spines or hairs at the apex of each segment.

Adult Characteristics. Eyes usually small and remote in both sexes, large in males of a few species. Mesothorax robust. Forewings more or less broad, cubito-anal area slightly to strongly expanded. Hind wings absent, or greatly reduced and with long and straight or recurved costal projection. Three caudal filaments present.

Haplohyphes Allen (figs. 335, 337, 380)

Nymphal Characteristics. Unknown, but presumably similar to *Leptohyphes*.
Nymphal Habitat. Unknown.
Nymphal Habits. Unknown.
Life History. Unknown.
Adult Characteristics. Wing length: 4.0–6.5 mm. Compound eyes of both sexes small. Forewings with forty to fifty-five crossveins behind vein R_1; veins CuP and A_1 strongly converge at wing margin. Hind wings with long recurved costal projection; hind wings present in both sexes. Fore tarsi about one-third length of tibiae. Male genitalia illustrated in fig. 380.
Mating Flights. Unknown.

Taxonomy. Allen, 1966b. *J. Kans. Entomol. Soc.* 39:566. Type species: *H. huallaga* Allen; type locality: Rio Huallaga, Tingo María, Huánuco Province, Peru.

Distribution. Neotropical. One species, *H. mithras* (Traver), is known from Costa Rica.

Leptohyphes Eaton (figs. 179, 180, 184–186, 336, 381, 418)

Nymphal Characteristics. Body length: 3–5 mm. Body somewhat flattened, usually more slender than *Tricorythodes*. Hind wing pads present in male, absent in female. Femora broad, with a transverse row of spines. Claws without denticles. Abdominal gills on segment 2 elongate, oval, always widest in apical half, operculate, covering following pairs. Illustration: Fig. 418.

Nymphal Habitat. The nymphs are found in large rivers in the United States and in a variety of streams and rivers in Mexico and Central America. They occur on submerged sticks, logs, branches, rocks, and vegetation.

Nymphal Habits. The nymphs are poor swimmers and usually crawl very slowly.

Life History. Unknown.

Adult Characteristics. Wing length: 3–8 mm. Eyes small, remote in most species; eyes large in males of some species. Forewings widest in region of MA–MP, cubitoanal area not expanded as in male of *Tricorythodes*; crossveins ranging in number from 43 to 136 behind vein R_1; vein CuP abruptly recurved, but not converging to or near vein A_1 at margin. Hind wings present in male, absent in female; two longitudinal veins commonly present, indistinct except near base; often a shorter vein runs upward from base toward costal angulation; costal angulation well developed, long, slender, and more or less recurved; subimaginal cuticle apparently not shed from hind wings. Forelegs shorter than body; fore tibiae three or four times length of tarsi. Male genitalia illustrated in fig. 381.

Mating Flights. Unknown.

Taxonomy. Eaton, 1882. *Entomol. Mon. Mag.* 18:208. Type species: *L. eximius* Eaton; type locality: Cordova, Argentina. Allen (1967) has described most of the North and Central American species from nymphs. No key to the species is available, but the nymphs should be identifiable from Allen's descriptions and drawings. We have deliberately emended the spelling of *L. murdocki* to *murdochi*; the species was named as a patronym of W. P. Murdoch (not Murdock).

Distribution. Neotropical and Nearctic. Widespread in Central America and Mexico with several species extending north to Utah, Texas, and Maryland.

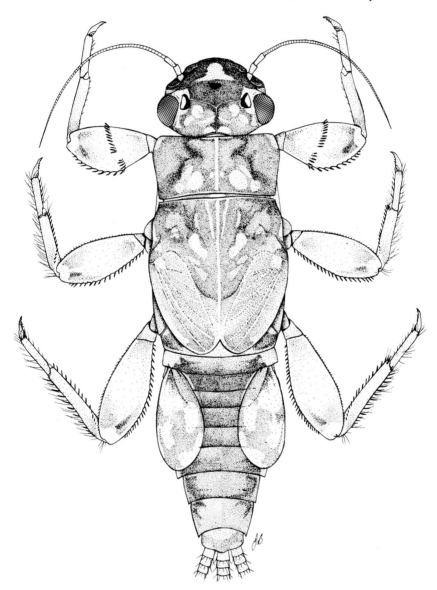

Fig. 418. *Leptohyphes packeri*, nymph.

Species	United States						Canada			Mexico		Central America
	SE	NE	C	SW	NW	A	E	C	W	N	S	
alleni Brusca											X	...
apache Allen	...	...	...	X	...	...	...	...	...	...		...
baumanni Kilgore & Allen	...	...	...	X	...	...	...	...	...	...		
berneri Traver	...						...	...	...	...	...	X
brevissimus Eaton	...								...	...	...	X
brunneus Allen & Brusca	...								...	...	X	...
castaneus Allen	...								...	...	...	X
consortis Allen & Brusca	...								...	...	X	
costaricanus Ulmer												X
dicinctus Allen & Brusca	...									...	X	...
dolani Allen	X	...	X	...								...
ferrugineus Allen & Brusca	...								...	...	X	
hispidus Allen & Brusca	...						...		...	...	X	...
lestes Allen & Brusca	...								...	...	X	...
lumas Allen & Brusca	...								...	...	X	
melanobranchus Allen & Brusca	...	...	...									X
mirus Allen	...	...	...	X								...
murdochi Allen												X
musseri Allen												X
nanus Allen												X
packeri Allen	...								X	X		X
phalarobranchus Kilgore & Allen	...	...	...	X	...							...
pilosus Allen & Brusca	...	...								...	X	...
priapus Traver	...						...	...	...	...	...	X
quercus Kilgore & Allen	...	...	...	X								...
robacki Allen	X	X										...
sabinas Traver	...	...	X	...						X		...
spiculatus Allen & Brusca	...	...								...	X	...
zalope Traver	...	...								...	X	...

Tricorythodes Ulmer (figs. 47, 51, 178, 181, 182, 219, 383, 419)

Nymphal Characteristics. Body length: 3–10 mm. Body short and stout. Hind wing pads absent. Femora moderate to broad with transverse or marginal rows of setae. Claws long and hooked at apex. Gills present on abdominal segment 2 triangular or subtriangular, always widest in basal one-third, operculate. Illustration: Fig. 419.

Nymphal Habitat. The nymphs occur in streams which have a perceptible current; they are found amid fine sand and gravel on the stream beds, in moss or other plant growth on large stones, and sometimes near the bases of rooted vegetation or among the exposed, washed roots of terrestrial plants. Most streams from which nymphs have been taken are permanent, but some are dry during winter. The streams are often silted, varying in size from rather small creeks to large rivers.

Nymphal Habits. Principally herbivorous. The nymphs are awkward swimmers, using an undulatory action of the abdomen to produce forward movement. They move almost entirely by crawling.

Life History. In Utah *Tricorythodes minutus* nymphs do not appear in most streams until late spring, the time for development determined as being less than seventeen weeks in one stream. The species seems to be multivoltine. R. J. Hall (doctoral dissertation, 1975) reports that in Minnesota *T. atratus* grows rapidly after a long egg diapause or an intermediate generation. The greatest density of nymphs exists in late spring and early summer, but the population increases slightly again in late summer and early fall, representing two summer generations in rapid succession. *T. atratus* eggs hatch continuously, and adults emerge every day during the warm months but in greatest numbers in July. The nymphs develop in about five weeks after hatching. Overwintering occurs in the egg stage. Hall observed subimagoes emerging underwater, the males beginning after sunset and continuing throughout the night and the females beginning the process at dawn and completing emergence within two hours. Exuviation to the imaginal stage occurs about an hour before dawn for the males and after dawn for the females. In other species, the nymph comes to the water surface, and the subimago immediately bursts free. At the time of emergence large numbers of subimagoes appear during a short time. Hall notes that in *T. atratus* the male subimaginal stage lasts from five to seven hours but that of the female is briefer, with molting occurring almost immediately or requiring up to two hours. In the southeastern part of the range, especially in Florida, emergence occurs from February to October; farther north the period of emergence ranges from May to late October. Some species of *Tricorythodes* are said to shed the subimaginal skin while in flight, but this point is in dispute. The adults of at least some species alight to molt.

Adult Characteristics. Wing length: 3–9 mm. Most species blackish, but some dark brown. Eyes small and widely separated in most species; in a few species eyes large in male. Pair of tubercles often present near posterior margin of vertex. Wings with relatively numerous crossveins; wings (especially of male) well developed in cubitoanal area; veins Sc and R generally

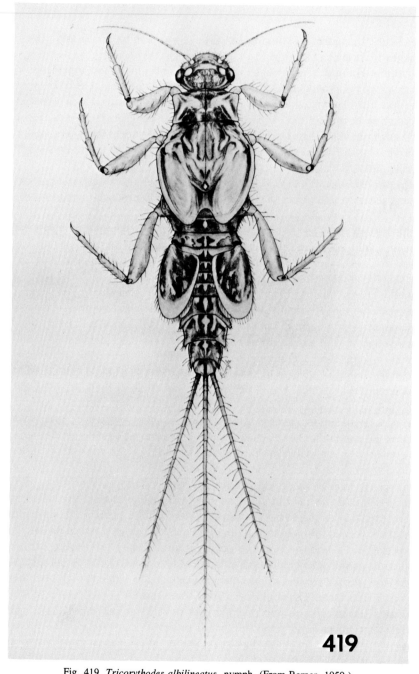

419

Fig. 419. *Tricorythodes albilineatus*, nymph. (From Berner, 1950.)

purplish black except at apex. Hind wings absent. Forelegs of male as long as body; fore tibiae 1½ times as long as fore tarsi; basal tarsal segment very short, second segment half length of tibiae and as long as tarsal segments 3–5 combined. Male genitalia illustrated in fig. 383.

Mating Flights. *T. allectus* has been observed in New York swarming in great numbers at midday over an open area above a bridge. Males of *T. minutus* fly rapidly in an up-and-down pattern over rivers at heights of five to fifty feet. The swarming of *T. minutus* reaches its greatest intensity from 9:30 A.M. to 10:30 A.M.; the flurry of some mass flights is so great that it prompted Needham to call them "snowflake mayflies." Their small size, their large numbers, and the difficulty of preparing and presenting suitable artificial flies resembling them have earned them the name also of "fisherman's curse." Leonard and Leonard (1962) report a Michigan species mating during the early morning hours from about 4:30 A.M. to 10:00 A.M., with death following very shortly after the eggs were laid. Hall observed males of *T. atratus* in Minnesota flying in large swarms from twenty-five to fifty feet above a stream just at sunrise. The females rose to meet the males; after mating the females descended to the water surface for oviposition. Small flights were seen as late as 10:00 A.M.

Taxonomy. Ulmer, 1920. *Stett. Ent. Zeit.* 81:122. Type species: *T. explicatus* (Eaton); type locality: North Sonora, Mexico. The adults of the species in this genus are particularly poorly known. If care is used, the adults of a number of the North American species can be identified by using the keys of Traver (in Needham, Traver, and Hsu, 1935) and Burks (1953). Allen (1967) has described additional species from the nymphs alone. Further rearing and the careful analysis of known species are needed to solve the taxonomic problems in this genus.

Edmunds and Allen (1957a) synonymized *T. fallacina* McDunnough with *T. minutus* Traver. After more study it now appears that the population named *T. fallacina* is most similar to *T. fallax* Traver. But *T. fallax* differs from *T. minutus* only in color; most notably, *T. fallax* has a distinct black ring at the base of each tibia. *T. fallax* and *T. minutus* intergrade completely along the eastern borders of California, Oregon, and Washington, but the black-ringed tibial character may extend farther eastward because McDunnough refers the northern populations to *T. fallax*. At most *T. fallax* and *T. minutus* seem worthy of subspecific rank, but we believe that these will best be regarded as mere clinal variants and hereby designate *fallax* as a synonym of *minutus*. It is probable that all records of *T. explicatus* (Eaton), except for those from Mexico, Texas, New Mexico, and Arizona, should be regarded as referable to *T. minutus*.

Distribution. Neotropical and Nearctic. Widespread in the Nearctic, extending as far north as British Columbia, Saskatchewan, and Quebec.

Species	United States						Canada			Mexico		Central America
	SE	NE	C	SW	NW	A	E	C	W	N	S	
albilineatus Berner	X											
allectus Needham		X	X									
angulatus Traver										X		
atratus McDunnough			X				X					
comus Traver									X			
condylus Allen				X								
corpulentus Kilgore & Allen				X								
dimorphus Allen				X								
edmundsi Allen				X						X		
explicatus (Eaton)	X		X							X		
fictus Traver			X									
minutus Traver		X		X	X		X	X				
mulaiki Traver										X		
notatus Allen & Brusca										X		
peridius Burks			X									
sordidus Allen											X	
stygiatus McDunnough	X		X				X					
texanus Traver			X									
ulmeri Allen & Brusca												X

Family NEOEPHEMERIDAE

This family is most closely allied to Caenidae, and the proto-Neoephemeridae were ancestral to the caenids. The two families constitute the Caenoidea, which in turn are allied closely to the Prosopistomatoidea.

The nymphs of Neoephemeridae are distinguishable from those of the Caenidae by only a few minor external characters. The rounded lobe on the anterolateral corners of the mesonotum which distinguishes North American nymphs of Neoephemeridae from those of Caenidae is not present in all Neoephemeridae, but all Neoephemeridae have hind wing pads and the fibrilliform portion of the gill on segments 3–6. The adults superficially resemble the Ephemeroidea, especially in their potamanthidlike wing venation.

The family is Holarctic and Oriental, with the greatest diversity in Asia. *Neoephemera* is the only Nearctic genus of the family. See *Neoephemera* for characteristics and distribution in North America.

Neoephemera McDunnough (figs. 48, 198, 384, 420)

Nymphal Characteristics. Body length: 8–17 mm. Body elongate. Head rounded, almost twice as wide as long; eyes lateral. Prothorax longer at margins than at midline. Mesonotum with distinct rounded lobe on anterolateral corners. Claws without denticles. Prominent median tubercle present on terga 1 and 2. Lateral margins of abdominal segments 3–9 prolonged posteriorly into flattened projections that may be decurved; exceptionally well developed on segments 6–9. Gills on abdominal segment 1 vestigial; operculate gills on segment 2 quadrate, fused at midline. Three caudal filaments with whorls of stiff setae. Illustration: Fig. 420.

Nymphal Habitat. In slow to moderately rapid streams. The nymphs of most species occur in debris and may also occur in numbers among exposed, well-washed roots of terrestrial plants. They are also frequently found among tangles of branches that have washed into the stream. Some species live deep within moss which is exposed to the current. *N. purpurea* sometimes occurs under large flat rocks in midstream where the flow of water is swift; the nymphs in such habitats find protected places where they encounter very little current.

Nymphal Habits. When dislodged, the nymphs immediately settle to the bottom and attach themselves to some object to which they cling tenaciously. They are awkward and slow when forced to swim. The caudal filaments are bent over the abdomen and lashed so that the movement of the abdomen, assisted by the beating of the almost bare caudal filaments, propels the insect.

Life History. The time required for development has not been determined, but evidence indicates that it is one year or less. When ready to emerge, the nymph rises to the surface, and the adult escapes. Occasionally the nymph crawls out of the water to emerge. In Florida emergence begins in March and extends into early May. Farther north, emergence is somewhat later with *N. purpurea* emerging in late June. In Canada *N. bicolor* has been recorded in July. The time of emergence is variable, with some subimagoes appearing at night, before dawn, or in the morning, and others appearing in the late afternoon or early evening.

Adult Characteristics. Wing length: 8–17 mm. Eyes of male large, separated dorsally by distance equal to about half the diameter of the eye. Forewings with basal costal crossveins weak or atrophied; stigmatic crossveins partly

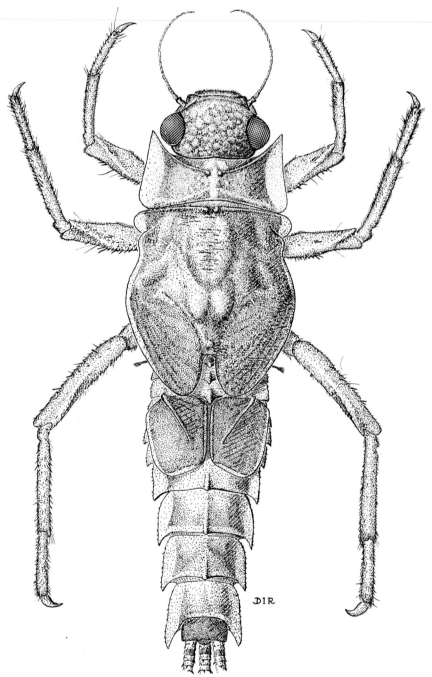

Fig. 420. *Neoephemera purpurea*, nymph. (From Edmunds, Allen, and Peters, 1963.)

anastomosed; vein MP_2 curved toward vein CuA near base; vein A_1 with from one to three crossveins attaching it to the hind margin. Hind wings with acute costal projection. Forelegs of male with tarsi subequal to tibiae; claws dissimilar. Pleural folds of abdomen expanded laterally. Male genitalia illustrated in fig. 384. Three well-developed caudal filaments.

Mating Flights. Not described.

Taxonomy. McDunnough, 1925. *Can. Entomol.* 57:168. Type species: *N. bicolor* McDunnough; type locality: Laprairie, Quebec. Berner (1956) has provided keys and numerous illustrations to separate the four known species of this genus in both adult and nymphal stages.

Distribution. Holarctic. In the Nearctic all species are eastern. One species occurs from Michigan to Quebec; others occur in the southeastern United States.

	United States						Canada			Mexico		Central
Species	SE	NE	C	SW	NW	A	E	C	W	N	S	America
bicolor												
McDunnough	...	...	X	...	...	...	X	...	...	...	...	...
compressa												
Berner	X	...	...	...	...	...	...	...	...	...	...	...
purpurea												
(Traver)	X	...	...	...	...	...	...	...	...	...	...	...
youngi Berner	X	...	...	...	...	...	...	...	...	...	...	...

Family CAENIDAE

This family of Caenoidea has a number of specialized characters in both adults and nymphs. The family is almost certainly derived from the proto-Neoephemeridae of the *Potamanthellus* group lineage. The nymphs of Caenidae are so similar to some Neoephemeridae that only the absence of hind wing pads and the absence of the fibrilliform tuft on the gills of segments 3–6 serve to distinguish the nymphs of Caenidae. The adults of Caenidae, however, are remarkably different from the ancestral pattern which persists in the Neoephemeridae.

The family occurs in a variety of ponds, lakes, streams, and rivers over most of the earth, being absent only from New Zealand and most oceanic islands.

Nymphal Characteristics. Thorax robust, abdomen moderately to strongly flattened; hind wing pads absent. Abdominal segments 3–6 shortened; quadrate gills on segment 2 operculate, not fused at midline; gills on segments 3–6 with margins fringed, fibrilliform tufts wanting. Three caudal filaments present.

Adult Characteristics. Eyes of both sexes small and remote. Thorax robust, abdomen relatively small and short. One pair of wings; veins MP_2 and IMP almost as long as vein MP_1; anal area of wings expanded, broad; hind margins of wings of imago with setae. Genital forceps of male one-segmented. Three caudal filaments, very long in male, short in female.

Brachycercus Curtis (figs. 328, 382, 421)

Nymphal Characteristics. Body length: 3–8 mm. Body flattened, broad. Head with three ocellar tubercles. Maxillary and labial palpi two-segmented. Forelegs relatively short; middle and hind legs longer; claws long and slender. Abdomen depressed; lateral margins of abdominal segments produced as broad, flat, bladelike upcurved projections. Illustration: Fig. 421.

Nymphal Habitat. In flowing streams the nymphs live close to the edge on sand with a very thin overburden of silt. They occur most frequently in water three or four inches deep, but they have been taken from water as much as twenty feet deep near sandy ridges formed on the stream bed. They also occur near the shores of lakes.

Nymphal Habits. When collected, the nymphs crawl so slowly that their progress is almost imperceptible; they raise the tail over the abdomen when they walk. In the natural habitat the nymphs are partially covered by silt and are very difficult to see.

Life History. The time required for nymphal development has not been determined. In the north emergence is restricted to the warm summer months, but longer emergence periods occur in the south. They may emerge throughout the year in the extreme southeastern part of the United States. Emergence in Idaho and Canada has been recorded as late as September. The adult stage is very brief, lasting only a few hours at most.

Adult Characteristics. Wing length: 3–6 mm. Antennal pedicel three times as long as scape. Prosternum half as long as broad, rectangular in shape; fore coxae widely separated. Abdomen small and contracted. Male genitalia illustrated in fig. 382.

Mating Flights. The subimagoes emerge in the early morning hours in some species and at nightfall in others, molting within minutes and probably mating soon after. Flights have been observed in early morning in Ontario, in mid-morning in northwest Florida and Wyoming, and after dark in Idaho.

Taxonomy. Curtis, 1834. *London and Edinburgh Philosophical Magazine*

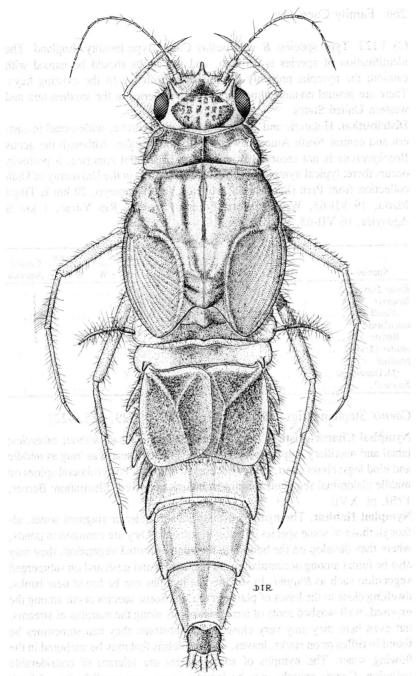

Fig. 421. *Brachycercus* sp., nymph. (From Edmunds, Allen, and Peters, 1963.)

(3) 3:122. Type species: *B. harrisellus* Curtis; type locality: England. The identification of species is difficult, and the adults should be named with caution; the nymphs probably cannot be identified with the existing keys. There are several unnamed species in North America in the southeastern and western United States.

Distribution. Holarctic and Neotropical. In the Nearctic, widespread in eastern and central North America west to Utah and Idaho. Although the genus *Brachycercus* is not recorded from Mexico or Central America, it probably occurs there; typical nymphs of the genus are present in the University of Utah collection from Peru (San Martin Province, Rio Tulumayo, 20 km E Tingo María, 19-VII-63, W. L. Peters; Loreto Province, Rio Yurac, 1 km S Aguaytía, 16-VII-63, W. L. Peters).

Species	United States						Canada			Mexico		Central America
	SE	NE	C	SW	NW	A	E	C	W	N	S	
flavus Traver	...	...	X	...	...	...	...	...	...	...	...	...
lacustris (Needham)	...	X	X	...	...	...	X	...	...	...	...	...
maculatus Berner	X	...	...	...	...	...	...	...	...	...	...	...
nitidus (Traver)	X	...	...	...	...	...	...	...	...	...	...	...
prudens (McDunnough)	...	...	X	...	...	...	...	X	...	...	...	...
Species?	...	...	...	X	X	...	...	...	...	...	...	...

Caenis Stephens (figs. 46, 49, 187, 215, 220, 329, 385, 422)

Nymphal Characteristics. Body length: 2–7 mm. Head without tubercles; labial and maxillary palpi three-segmented. Forelegs nearly as long as middle and hind legs; claws short, somewhat curved apically. Posterolateral spines on middle abdominal segments prominent but not upcurved. Illustration: Berner, 1950, pl. XVII.

Nymphal Habitat. The nymphs usually inhabit quiet or stagnant water, although those of some species develop in streams. They are common in ponds, where they develop on the bottom in the zone of rooted vegetation; they may also be found among accumulations of leaf debris and trash and on submerged vegetation such as *Ruppia*. In streams the nymphs can be found near banks, dwelling close to the bases of plants in the silt. Some species occur among the exposed, well-washed roots of terrestrial plants along the margins of streams, but even here they stay very close to the substrate; they can sometimes be found in riffles or on sticks, leaves, or other debris that may be anchored in the flowing water. The nymphs of some species are tolerant of considerable pollution. *Caenis* nymphs may be found in even very small bodies of fresh water and at the edges of lakes which are overgrown with vegetation.

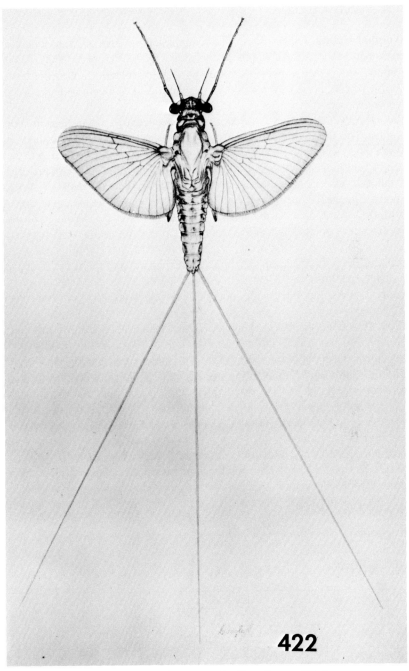

422

Fig. 422. *Caenis diminuta*, male. (From Berner, 1950.)

Nymphal Habits. Omnivorous. The nymphs feed chiefly on plant material but sometimes on dead nymphs or on other organic material. When taken from the water, they crawl slowly with a wriggling motion. Some have been observed to be positively phototropic, gathering in numbers at underwater light traps.

Life History. In Florida *C. diminuta* has been reared from eggs in four months. The time required for hatching the eggs varied from five days to eleven days. Apparently the length of the development period is related to the time of year in which the eggs are laid. There is some indication that the eggs of fall and winter broods hatch much more rapidly than do those of spring broods. Emergence occurs in the afternoon or at night. When ready to transform, the nymph floats at the surface in shallow water; the subimago bursts free from the nymphal skin and takes flight. Coming to rest on some support, the subimago molts to the imago stage five to six minutes after emergence. The adult lives for three or four hours. Subimagoes and adults are strongly positively phototropic. In the northern part of the range emergence occurs chiefly in June and July, with sporadic emergences occurring in May and September.

Adult Characteristics. Wing length: 2–5 mm. Antennal pedicel twice as long as scape. Prosternum two or three times as long as broad, triangular in shape; fore coxae closely approximated. Male genitalia illustrated in fig. 385.

Mating Flights. In Florida mating was observed in late afternoon about five feet above the surface of the water. The adults swarm within a few minutes after emerging; they live only a few hours at most. *C. simulans* swarms in late evening; its flight consists of rapid vertical movements six to ten feet above the water or the ground.

Taxonomy. Stephens, 1835. *Illustrations of British Entomology* 6:61. Type species: *C. macura* Stephens; type locality: London, England. Although the species of this genus are difficult to separate, the adults should be identifiable with the existing keys. There are no satisfactory keys for identifying the nymphs.

Distribution. Cosmopolitan with the exception of the Australian realm and most oceanic islands. In the Nearctic, widespread throughout North America south to Panama.

	United States						Canada			Mexico		Central America
Species	SE	NE	C	SW	NW	A	E	C	W	N	S	
amica Hagen	X	X	...	...	...	...	...	...	...	...	...	...
anceps Traver	X	X	X	...	...	...	...	...	...	...	...	...
delicata Traver	...	...	X	...	...	...	...	...	...	...	...	...
diminuta Walker	X	...	...	...	...	...	X	...	...	...	...	...

Species	United States						Canada			Mexico		Central America
	SE	NE	C	SW	NW	A	E	C	W	N	S	
forcipata McDunnough	. . .	X	X	. . .	. . .	. . .	X	X	. . .	. . .	. . .	. . .
gigas Burks	. . .	. . .	X	. . .	. . .	. . .	. . .	. . .	. . .	. . .		. . .
hilaris (Say)	X	X	X	. . .	. . .	. . .	. . .	. . .	. . .	. . .	. . .	. . .
jocosa McDunnough	X	X	X	. . .	. . .	. . .	X	. . .	. . .	. . .	. . .	. . .
latipennis Banks	. . .	X	. . .	. . .	X	. . .	X	X	. . .	. . .	. . .	. . .
punctata McDunnough	. . .	X	. . .	. . .	. . .	. . .	X	. . .	. . .	. . .	. . .	. . .
ridens McDunnough	X	. . .	X	. . .	. . .	. . .	X	. . .	. . .	. . .	. . .	. . .
simulans McDunnough	. . .	X	X	X	X	. . .	X	X	X	. . .	. . .	. . .
tardata McDunnough	. . .	. . .	. . .	. . .	. . .	. . .	. . .	X	X	. . .	. . .	. . .
Species?	. . .	. . .	. . .	. . .	. . .	. . .	. . .	. . .	. . .	. . .	. . .	X

Family BAETISCIDAE

The family Baetiscidae contains only the genus *Baetisca*. The family is an isolated one whose nearest relative is the Old World genus *Prosopistoma*, the only genus of Prosopistomatidae. The two families are placed in the Prosopistomatoidea, a superfamily believed to have been derived from ancestral pre-Caenoidea.

The family is endemic in North America. See *Baetisca* for characteristics and distribution.

Baetisca Walsh (figs. 26, 226, 387, 423)

Nymphal Characteristics. Body length: 4–14 mm. Body stout. Head frequently with prominent frontal projections and spinous projections from anterior of genae. Large mesonotal shield, or carapace, completely covering dorsum of mesothorax, metathorax, and abdominal terga 1–5 and basal half of tergum 6; apical margin fits into pyramidal elevation on tergum 6; shield often with prominent lateral spines. Legs short; claws rather long and slender, without denticles. Gills present on segments 2–6, enclosed in gill chamber covered by shield. Posterolateral spines usually well developed on abdominal segments 6–9. Three caudal filaments; short, approximately equal in length. Illustration: Fig. 423.

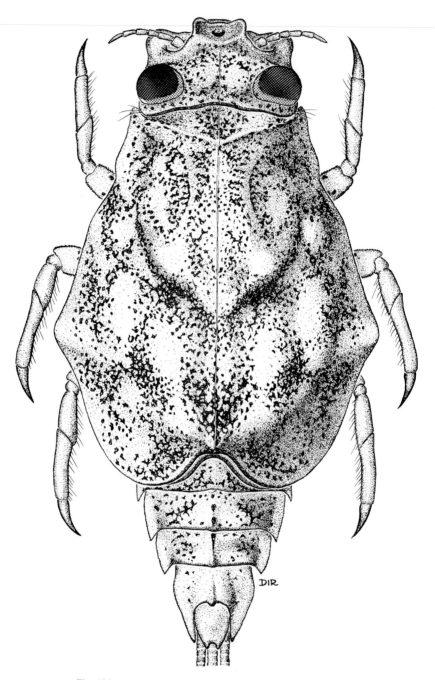

Fig. 423. *Baetisca columbiana*, nymph. (From Edmunds, 1960.)

Nymphal Habitat. Usually in small to medium-sized streams, occasionally in large rivers and along the edges of lakes where there is some wave action. The nymphs live in a variety of habitats; most sit on top of the sand in the lee of rocks or gravel or in pebbly riffles where they are somewhat protected. One species occurs in quiet waters near or at the banks of slow-flowing streams. Nymphs of other species partially bury themselves in fine gravel and sand or in a mixture of silt and sand. Lehmkuhl (1972) reports that *B. obesa* avoids silt and crawls on rocks or submerged grass. The nymphs are chiefly herbivores and detritivores.

Nymphal Habits. When quiet, the nymphs often hold the caudal filaments over the abdomen. They swim by pulling the legs under the body, drawing together and depressing the tail, and rapidly vibrating the posterior portion of the body. They swim in short spurts; when they come to rest, they spread the legs and seize some supporting object. Some species rest on the sand with the body partially covered, thus becoming difficult to detect against the background.

Adult Characteristics. Wing length: 8–12 mm. Eyes of male large, almost contiguous dorsally. Thorax robust; posterior margin of prosternum with two spinous projections between fore coxae. Wings sometimes washed with red, orange, amber, or violet, at least basally; numerous weak crossveins in forewing; many short marginal intercalaries; outer wing margin slightly scalloped. Forelegs of male almost as long as body; tarsi more than twice length of tibia; basal tarsal segment about 1½ times as long as second segment; claws of fore tarsi similar, blunt; those of middle and hind legs dissimilar. Abdomen broad basally, rapidly tapering posteriorly. Male genitalia illustrated in fig. 387. Two caudal filaments, usually shorter than the body; terminal filament vestigial.

Mating Flights. Several species have been seen swarming in midmorning or late morning. The swarms were five to ten feet above the water. *B. escambiensis* swarms in late November in Florida. Despite special efforts, no swarms of *B. rogersi* have been located (Peters, personal communication; Pescador and Peters, 1974).

Life History. The southeastern *B. rogersi* has been studied in detail (Pescador and Peters, 1974). The first eggs from fertilized females hatch at an average of 21 days after oviposition; about 1% of the eggs from unfertilized females also develop, but the first eggs hatch at an average of 26.6 days after oviposition. Nymphal development requires twelve instars with a four-month interval between the first and twelfth instars. The young nymphs appear in September and are present through June. The emergence of adults begins in March and extends through early July, with the peak in April. There is a single generation each year. When ready to emerge, the nymphs crawl from the water onto any object protruding above the surface. The subimagoes appear between 8:30 A.M. and 2:30 P.M., with the peak of emergence between 8:30 A.M. and 10:30 A.M. In Michigan emergence occurs in late May and early July. *B. bajkovi* eggs hatch in late summer, and the nymphs pass the winter under the ice (Lehmkuhl, 1972). The adults emerge the following June or July.

Taxonomy. Walsh, 1862. *Proc. Acad. Natur. Sci. Philadelphia* 13–14:378. Type species: *B. obesa* (Say); type locality: Rock Island, Illinois. Specific identification generally should be possible in this genus, although the characters of the adults are not always clearly defined. There should be little difficulty in assigning the nymphs to species.

Distribution. Nearctic. The genus is most abundant and diverse in the southeastern United States. Several species also occur in eastern and central North America. Single nymphs have been collected in Wyoming and Washington, and there is a highly dubious record of *B. obesa* from California.

Species	United States						Canada			Mexico		Central America
	SE	NE	C	SW	NW	A	E	C	W	N	S	
bajkovi Neave	...	...	X	...	...	...	...	X	X	...	...	...
becki Schneider & Berner	X											
callosa Traver	X	X					X					
carolina Traver	X						X					
columbiana Edmunds							X					
escambiensis Berner	X											
gibbera Berner	X											
lacustris McDunnough	...	...	X	...	...	...	X	X	...	...		
laurentina McDunnough		X	X									
obesa (Say)	X	X	X				X					
rogersi Berner	X											
rubescens (Provancher)							X					

Family PROSOPISTOMATIDAE

The Prosopistomatidae (superfamily Prosopistomatoidea) and Baetiscidae probably arose from a common ancestor that was a pre-caenoid. The Caenoidea constitute another branch of this phyletic line. It seems unlikely that the highly specialized carapace of the Prosopistomatoidea nymphs would have evolved from nymphs which already have a modification for gill protection in the form of the specialized quadrate gill 2 of the Caenoidea.

The family is recognized for a single genus that is widespread in the Old World tropics, with one species in Europe. The genus is found in Africa, Madagascar, Ceylon, the Sunda Shelf, the Philippines, New Guinea, and Europe. The family is unknown in the western hemisphere.

Nymphal Characteristics. Head recessed into the thorax. Pronotum and mesonotum fused to form a shield extending to abdominal segment 6. Legs small, not visible dorsally. Caudal filaments very short, retractable.

Adult Characteristics. Eyes of both sexes small, widely separated. Thorax robust. Forewings without crossveins; in disc of wing, longitudinal veins of male with an intercalary on each side about half as long as vein; forewing of female with eleven or fewer veins behind R_1. Forelegs of male barely longer than middle and hind legs; legs of female vestigial. Caudal filaments of male about three-fifths length of forewings; caudal filaments of female about one-tenth length of forewings. The females do not molt to the imago stage.

Family BEHNINGIIDAE

This family is an isolated and primitive phyletic line within the superfamily Ephemeroidea. It seems unlikely that the Behningiidae could have evolved from any of the true burrowing nymphs that have the forelegs modified for digging or even from the nonburrowing Euthyplociidae and Potamanthidae with large mandibular tusks. The ancestral precursor of the Behningiidae must have had a nymph in which the tusks were absent or small, the forelegs were not modified for burrowing, and the gills on segment 1 were single. The ancestral nymph could have resembled the European *Potamanthus luteus* Linnaeus. The nymphs are unique among Ephemeroidea in that the mandibles lack tusks. The middle and hind legs are modified as protective devices for the gills. The forelegs are palplike. The gills are ventral, a unique feature within the Ephemeroidea, but otherwise typical of the superfamily except that the single gills on segment 1 are longer than those on the other segments.

The family is Holarctic in distribution with one genus ranging from Poland to Siberia, a second genus in Siberia, and the genus *Dolania* in North America. See *Dolania* for characteristics and distribution in North America.

Dolania Edmunds and Traver (figs. 42, 197, 202, 204, 424)

Nymphal Characteristics. Body length: 13 mm. Body rounded. Head flattened; antennae inserted ventrally; anterior margin with a pair of patches of spines and setae; labrum emarginate medially; mandibles small, without mandibular tusks. Anterolateral corners of pronotum produced and crowned with spines, posterolateral corners produced. Prothoracic legs palplike; mesothoracic legs highly modified, tibiae and tarsi forming a spinous pad; metathoracic legs highly modified; all legs without claws. Abdominal terga with dense, laterally extended setae and a row of setae on the posterior margin of each segment; sterna with rows of long setae. Gills on segments 1–7; gill 1 single, others bifid. Three caudal filaments. Illustrations: Fig. 424; McCafferty, 1975b, fig. 1.

Nymphal Habitat. Large rivers or tributaries with sandy bottoms. The nymphs are adapted for burrowing in sand. The usual habitat is moderately clean sand in fairly swift current.

Nymphal Habits. Nymphs that have been removed from the sand swim downward and quickly burrow straight down into the sand, disappearing almost immediately. When burrowing, the nymphs have no dorsal-ventral orientation. Forward motion is effected by the burrowing action of the forelegs and the elongated labial and maxillary palpi. The head functions as a bulldozer blade and also protects the mouthparts and the eyes. The thoracic prolongations and the drawn up, chelipodlike middle legs box in the bulky ventral anterior half of the nymph, thus increasing its efficiency as a burrower. The trailing hind tibiae and tarsi lie just below the ventral rows of gills and the elongate first gill is held motionless at the sides, thus the gills are free of sand at all times. Gills 2–7 move rhythmically to produce a constant current.

Life History. The adults emerge in Florida in late April and early May. By November the nymphs are about one-third grown. The life cycle takes approximately one year. The eggs of *Dolania* and other Behningiidae are as much as 1 mm in diameter, by far the largest known in mayflies.

Adult Characteristics. Wing length: 13–16 mm. Body light purplish brown, abdomen pale. Antennae inserted on anterolateral projections of head in both sexes, but obviously so in female. Veins Rs and MA forked about one-tenth the distance from base to wing margin; Rs forked nearer base than is MA; veins MP and CuA originate from vein CuP; four long intercalaries between CuA and CuP. All legs of both sexes feeble and more or less twisted. Penes of male twice as long as genital forceps. Male genitalia illustrated in fig. 204. Terminal filament much shorter and thinner than cerci.

Mating Flights. Not described in the literature presently available. W. L. and J. G. Peters have a detailed study of the species in progress.

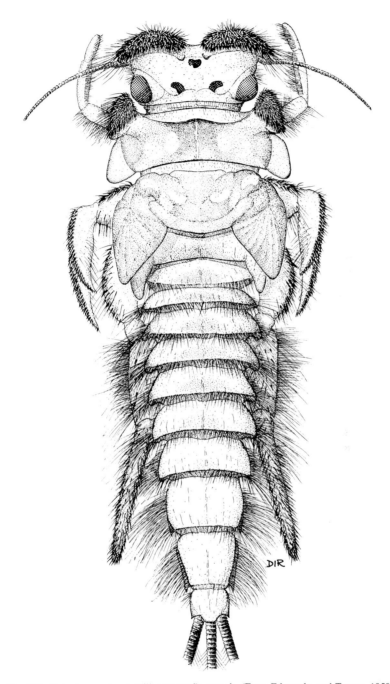

Fig. 424. *Dolania americana* (gills removed), nymph. (From Edmunds, and Traver, 1959.)

Taxonomy. Edmunds and Traver, 1959. *Ann. Entomol. Soc. Amer.* 52:46. Type species: *D. americana* Edmunds and Traver; type locality: twenty-five miles south of Aiken, Upper Three Runs, South Carolina. There is a single species in the genus.

Distribution. Nearctic. South Carolina, south to Florida. The genus is known from only a few scattered localities.

Family POTAMANTHIDAE

Potamanthidae is one of the most primitive families of the Ephemeroidea, a superfamily most probably derived from leptophlebiidlike ancestors. The family is most diverse in the Oriental realm and in Palearctic Asia. The genus *Potamanthus*, which is the only genus of the family in North America, is widely distributed in the Holarctic. See genus *Potamanthus* for characteristics and distribution in North America.

Potamanthus Pictet (figs. 32, 208, 217, 425)

Nymphal Characteristics. Body length: 8–15 mm. Nymph of sprawling type. Frontal margin of head somewhat rounded. Mandibular tusks projecting forward and visible from above; with only inconspicuous setae and short spines on dorsal surface and lateral margins. Femora flattened; tibiae and tarsi slender, cylindrical; long, curved spine at apex of fore tibiae. Gills lateral in position; gills on segment 1 rudimentary and unbranched; gills on segments 2–7 paired, each lamella slender and pointed, margins with long fringes. Three caudal filaments; bare at extreme base and at tip, terminal filament with setae on both sides, cerci with setae on mesal side only. Illustration: Fig. 425.

Nymphal Habitat. The nymphs live on the bottom amid silt, sand, and gravel or on stones in rather swiftly flowing shallow water.

Nymphal Habits. Not described.

Life History. The time required for development is probably one year, although it has not been determined definitely. Eggs have been hatched in the laboratory in fourteen days. Adults have been collected from May to mid-August.

Adult Characteristics. Wing length: 7–13 mm. Pale mayflies with vertex and dorsum of thorax light reddish brown. Eyes of male remote, separated by distance equal to at least one diameter of the eye. Forewings with basal part of

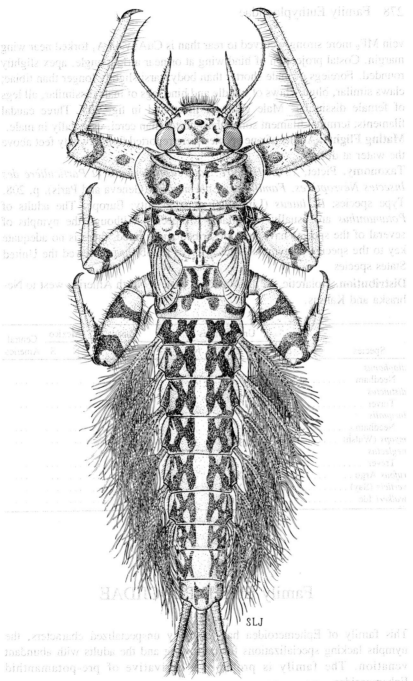

SLJ

Fig. 425. *Potamanthus* sp., nymph. (From Edmunds, Allen, and Peters, 1963.)

vein MP_2 more strongly curved to rear than is CuA; vein A_1 forked near wing margin. Costal projection of hind wing at or near a right angle, apex slightly rounded. Forelegs of male shorter than body; tarsi slightly longer than tibiae; claws similar, blunt; claws of middle and hind legs of male dissimilar, all legs of female dissimilar. Male genitalia illustrated in fig. 208. Three caudal filaments; terminal filament somewhat shorter than cerci, especially in male.

Mating Flights. Adults of one species were reported flying twenty feet above the water at dusk.

Taxonomy. Pictet, 1843. *Histoire Naturelle, Générale et Particulière des Insectes Névroptères. Famille des Ephémérides* (Geneva and Paris), p. 208. Type species: *P. luteus* (Linnaeus); type locality: Europe. The adults of *Potamanthus* are usually identifiable to species. Although the nymphs of several of the species have been described and figured, there is no adequate key to the species of nymphs. McCafferty (1975b) has reviewed the United States species.

Distribution. Holarctic. In the Nearctic, eastern North America west to Nebraska and Kansas.

	United States						Canada			Mexico		Central
Species	SE	NE	C	SW	NW	A	E	C	W	N	S	America
diaphanus Needham	...	X	...	...	...	...	...	...	...	...	...	...
distinctus Traver	X	X	X	...	...	...	...	...	...	...	...	...
inequalis Needham	...	X	...	...	...	...	...	...	...	...	...	...
myops (Walsh)	X	...	X	...	...	...	...	...	...	...	...	...
neglectus Traver	X	X	...	...	...	...	...	...	...	...	...	...
rufous Argo	X	X	X	...	...	...	X	...	...	...	...	...
verticis (Say)	X	X	X	...	...	...	X	...	...	...	...	...
walkeri Ide	...	X	...	...	...	...	X	...	...	...	...	...

Family EUTHYPLOCIIDAE

This family of Ephemeroidea has relatively unspecialized characters, the nymphs lacking specializations for burrowing and the adults with abundant venation. The family is probably a derivative of pre-potamanthid Ephemeroidea.

The family is of pantropical distribution with three of the seven genera occurring in South America. Of these, two genera are found in Central America and southern Mexico.

Nymphal Characteristics. Mandibular tusks longer than head and with numerous long hairs. Femora flat and broad, tibiae and tarsi cylindrical, not modified for burrowing. Gills on abdominal segment 1 small and ovoid. Gills on segments 2–6 extend laterally from abdomen, fringes longer than width of lamella. Three caudal filaments with only inconspicuous setae.

Adult Characteristics. Eyes of both sexes somewhat similar in size, rather small. Ocelli relatively large. Forewings nearly half as broad as long. Vein R_1 forked less than one-sixth distance from base; MA forked near or beyond fork of R_1. Cubital region with eight or more sigmoid veins attaching CuA to hind margin. Legs well developed. Females 1½ to 2 times as large as male; wings of female with more crossveins than those of male. Three caudal filaments.

Campylocia Needham and Murphy (figs. 43, 189, 201, 326, 426)

Nymphal Characteristics. Body length: 12–35 mm. Antennae about three-fourths length of tusks when fully mature, longer in young nymphs. Tibial spine of forelegs about one-fourth length of tarsi; claws of fore tarsi inserted at apex. Illustration: Fig. 426.

Nymphal Habitat. On large rocks in moderately rapid to swift rivers and streams.

Nymphal Habits. Unknown.

Life History. Poorly known. Adults have been collected during all months of the year except February, May, and June. In Peru emergence of all sub-imagoes from one stream apparently occurred within a few days.

Adult Characteristics. Wing length: 10–40 mm. Cubital area of forewings with one or two intercalaries paralleling vein CuA; these parallel intercalaries attached to hind margin of wing by about six to eight sigmoid intercalaries. Genital forceps without a short terminal segment. Male genitalia illustrated in fig. 326.

Mating Flights. Unknown.

Taxonomy. Needham and Murphy, 1924. *Bull. Lloyd Lib.* 24, Entomol. Ser. 4:25. Type species: *C. anceps* (Eaton); type locality: Rio Mauhes, Brazil. Only *C. anceps* occurs from Panama northward to Honduras.

Distribution. Principally tropical America, from Brazil and Peru north to Honduras. In view of the known habitats, the genus will probably be found to extend farther north than *Euthyplocia*.

Euthyplocia Eaton (figs. 188, 327, 338)

Nymphal Characteristics. Body length: 25–35 mm. Antennae three times as

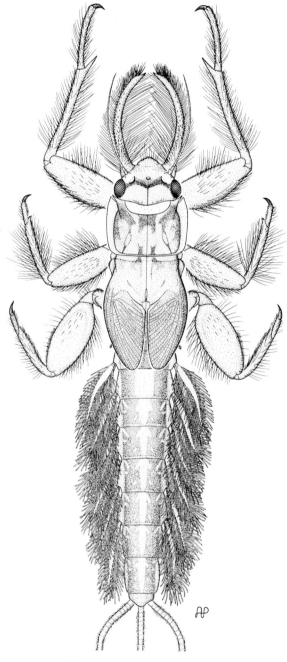

Fig. 426. *Campylocia* sp., nymph.

long as tusks. Tibial spine on forelegs long, more than half the length of tarsi; claws of fore tarsi inserted laterally, tarsi extend beyond base of claw. Illustration: Needham, 1920, pl. 79; same illustration in Needham and Needham, 1962, pl. IXA, fig. 1.

Nymphal Habitat. The nymphs occur in rivers on the surfaces of relatively large rocks and boulders. In Honduras the immature nymphs occur on rocky bottoms in rapid currents, where water is from one to two feet deep, while the mature nymphs occur on rocky bottoms in moderate currents where the water is only six to eight inches deep.

Nymphal Habits. Not described.

Life History. The adults are known to emerge in July.

Adult Characteristics. Wing length: 20–38 mm. Vein CuA of forewings connected to hind margin by ten to thirteen sigmoid intercalaries. Genital forceps with short terminal segment. Male genitalia illustrated in fig. 327.

Mating Flights. Not described.

Taxonomy. Eaton, 1871. *Trans. Entomol. Soc. London*, p. 67. Type species: *E. hecuba* (Hagen); type locality: Vera Cruz, Mexico. *E. hecuba* occurs north to Vera Cruz.

Distribution. The genus occurs principally in tropical America from Brazil and Peru north to Vera Cruz. *E. bullocki* Navas, described from Chile, is actually a species of the siphlonurid genus *Chaquihua*.

Family EPHEMERIDAE

The Ephemeridae are Ephemeroidea that probably arose from pre-potamanthid ancestors. The family includes two subfamilies. The Old World Palingeniidae presumably were derived from proto-*Pentagenia* of the Ephemeridae. McCafferty (1972) has placed *Pentagenia* in a separate family, Pentageniidae. We have recognized this proposed change in the classification to the extent of placing *Pentagenia* in a separate subfamily. *Pentagenia* nymphs seem to be at the palingeniid grade of evolution, but the adults remain undifferentiated from the ephemerids. Comparisons of *Pentagenia* with primitive members of the Palingeniidae are needed to clarify the taxonomic relationships in this group. It may be that the Pentageniinae are more properly placed in the Palingeniidae (W. P. McCafferty, personal communication). The family is widespread on all continents except Australia and is present in

New Zealand and Madagascar. Most of the species are large and conspicuous.
Nymphal Characteristics. Mandibular tusks well developed, projecting in front of head; tusks upcurved distally. Gills dorsal; gills on abdominal segment 1 reduced to vestiges, gills on segments 2–7 forked with margins fringed. Legs fossorial; hind tibiae terminating in an acute point. Three caudal filaments.

Adult Characteristics. Eyes of male moderate to large. All legs of both sexes functional and well developed; claws of middle and hind legs of male, and all legs of female, dissimilar, one sharp and one blunt. Forewings with veins MP_2 and CuA strongly divergent from vein MP_1 basally; vein A_1 unforked, attached to hind margin by three to many veinlets. Genital forceps of male with two long basal segments and one or two short apical segments. Two or three caudal filaments in both sexes.

Subfamily Ephemerinae

Nymphal Characteristics. Frontal process of head variable. Mandibular tusks without row of spurs along lateral margin. Galea-lacinia of maxillae two or more times as long as wide. Maxillary palpi relatively narrow, segments 2 and 3 more than four times as long as broad.

Adult Characteristics. Pronotum of male well developed, no more than twice as wide as long. Vein A_1 attached to hind margin by two to many veinlets. Tarsi of forelegs over three-fourths length of femora. Penes of male variable, not long and tubular. Caudal filaments of female longer than body.

Ephemera Linnaeus (figs. 37, 190, 210, 339, 386)

Nymphal Characteristics. Body length: 12–20 mm. Mandibular tusks with long setae principally in the basal half; with scattered spines. Head with frontal process bifid; usually widest near apex. Flagella of antennae with whorls of setae at least in basal two-thirds. Gills on segment 1 bifid near base. Anterior (dorsal) lamella of gill 2 with outer margin of platelike portion expanded only slightly at base. Illustration: Eaton, 1883–88, pl. 30.

Nymphal Habitat. In streams and lakes the nymphs burrow in sand with a small admixture of silt. In streams they are more common in quiet areas where the current is reduced. Stream habitats vary from small creeks to large rivers, depending on the species. In lakes the nymphs usually live in water not more than a few meters deep, but they are occasionally found in deeper water.

Nymphal Habits. When the nymphs are taken from their positions in the sand, they immediately begin burrowing. They exhibit very pronounced negative phototropism. The gills are moved in an undulating fashion, the movement beginning at the anterior end of the body and progressing toward the posterior end. The nymphs are fair swimmers, using undulating movements of the entire abdomen. The food of the nymphs apparently varies by species. Coffman et al. (1971) report *E. simulans* as being largely carnivorous and *E. varia* as being herbivorous.

Life History. The life history of only one species, *E. simulans*, has been determined. At Lake Wawasee, Indiana, the main emergence occurred during the last of May and the first few days of June. The eggs hatched within fourteen days in the laboratory. It is estimated that under natural conditions from twenty to thirty days are required for development. On hatching, the young nymphs are negatively phototropic and positively thigmotropic, and they begin burrowing immediately. Growth of the nymphs during the summer months is rapid, size increasing approximately 7 mm in thirty-nine days. The growth is variable and slows as the water cools, but the nymphs are probably almost mature before winter arrives. On the basis of evidence gathered at Lake Wawasee and Lake Erie, it appears that the life cycle requires one year. Emergence in various parts of North America occurs from May to August and rarely as late as October.

Adult Characteristics. Wing length: 10–15 mm. Frons does not extend below eyes. Eyes of male separated by a distance somewhat less than width of one eye. Crossveins of forewings crowded together at bullae; three to nine veins connect vein A_1 to wing margin; wings more or less dark spotted. Abdomen with conspicuous dark color pattern in most species. Genital forceps four-segmented; penes short, fused, and with a pair of ventral spines. Male genitalia illustrated in fig. 386. Terminal filament as long as cerci in both sexes.

Mating Flights. Males predominate in swarms over shrubs or trees along the shore of lakes or streams, flying in the usual up-and-down manner. The downward portion of the flight, from one to twenty feet, is accomplished with wings, legs, and tails outstretched. The up-and-down motion continues until the females enter the swarm, at which time the males approach and seize females from below, and copulation follows. Mating continues for less than a minute in flight, but the pairs lose altitude and sometimes strike the water surface or the ground before separating. Usually the male releases the female while the pair is still several feet in the air. The male then returns to the swarm, and the female flies over the water and releases her eggs. Mating usually takes place in the air, but adults of *E. guttulata* have been observed copulating while resting on the underside of a leaf.

Mating flights of *E. simulans* begin about sunset, and the males continue flying so late that they cannot be seen in the available light. Apparently with darkness the mating flights cease. The females die after oviposition, and it is believed that the males are spent by the end of the evening mating flight.

Adults of *E. guttulata* have been observed swarming over a road paralleling the St. Lawrence River at 11:00 A.M. on a very cloudy day.

Taxonomy. Linnaeus, 1758. *Systema Naturae*, 10th ed., p. 546. Type species: *E. vulgata* Linnaeus; type locality: Europe. The adults of this genus should be identifiable with the aid of keys and descriptions. The mature nymphs often show the adult body color pattern and sometimes the color pattern of the adult wing. *Ephemera simulans* is by far the most common and widespread species. The genus *Ephemera* has been divided into subgenera by McCafferty and Edmunds (1973), but only the nominal subgenus occurs in North America. McCafferty (1975b) has reviewed the genus *Ephemera* in the United States.

Distribution. Ethiopian, Oriental, and Holarctic. The genus is most abundant and diverse in Asia. In the Nearctic, south to Utah, Colorado, Oklahoma, and Florida.

Species	United States						Canada			Mexico		Central America
	SE	NE	C	SW	NW	A	E	C	W	N	S	
blanda Traver	X	...	...	...	...	...	...	...	...	...	...	...
compar Hagen	...	...	...	X	...	...	...	...	...	...	...	...
guttulata												
Pictet	X	X	...	...	...	...	X	...	...	...	...	...
simulans												
Walker	X	X	X	X	X	...	X	X	X	...	...	...
traverae Spieth	...	...	X	...	...	...	...	...	...	...	...	...
triplex Traver	...	X	...	...	...	...	...	...	...	...	...	...
varia Eaton	X	X	X	...	...	...	X	...	...	...	...	...

Hexagenia Walsh

Nymphal Characteristics. Body length: 12–32 mm. Mandibular tusks with long setae on mesal and lateral margins. Head with frontal process widest at base, apex truncate, conical, or rounded. Flagella of antennae with whorls of long setae in basal two-thirds. Gills on segment 1 very small, bifid near base. Anterior (dorsal) section of gills on segment 2 with basal portion distinctly expanded in basal half. Illustration: Fig. 427.

Nymphal Habitat. Vast numbers of nymphs occur in streams and lakes, where the nymphs burrow into the soft bottom. In streams they inhabit quiet back waters with an accumulation of silt. In lakes they occur at considerable depths. Depth in itself does not appear to have a great influence on distribution; the nymphs are found in very shallow water and down to depths of about eighteen meters, but they are less abundant in deeper waters. The maximum depth apparently occurs in but not below the thermocline. The U-shaped burrows may occur in large numbers, and the openings can be easily seen in

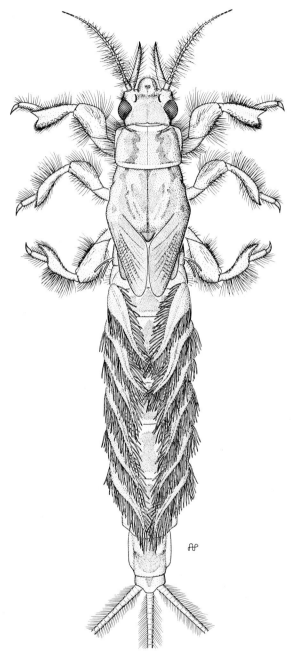

Fig. 427. *Hexagenia* (*Hexagenia*) *limbata*, nymph.

shallow water. The character of the bottom is of great importance in determining local distribution. The nymphs are usually found in large numbers only in bottoms composed of soft mud or clay, and they are largely restricted to a substratum which is soft, yet firm enough to permit the maintenance of a burrow. They do not ordinarily inhabit sand, gravel, rubble, or peat bottoms or bottoms which are flocculent. Burrowing to a depth of about five inches, the nymphs maintain the flow of water through the tunnel by rhythmically waving the gills. The nymphs appear to be unable to withstand stagnation in which the dissolved oxygen content of the water falls below 1 ppm.

Nymphal Habits. Swimming is accomplished by dorsoventral undulations of the body, chiefly the abdomen. The caudal filaments are directed backward and overlap one another. The forelegs and middle legs are directed anteriorly and the hind legs posteriorly, all legs being held close to the body. The gills, extended up and laterally, are waved continuously in swimming. Disturbed nymphs usually swim vigorously toward the bottom and attempt to burrow into it. Burrowing is accomplished in seconds. The nymphs enter the mud by digging with the forelegs and pushing with the hind legs, aided occasionally by a strong undulating movement of the abdominal gills. Once the head and the legs are under the mud, the remainder of the body quickly disappears. The nymphs are largely filter feeders, but they leave the burrow to feed on the mud surface. In rivers a small percentage of the population drifts downstream.

Life History. In Michigan *H. limbata* probably requires one year to develop, with the rate of growth being a function of the water temperature. In Utah *H. limbata* emerges from canal channels that were dry seventeen weeks earlier. Under laboratory conditions of high temperature some adults of *H. bilineata* were reared from eggs in about thirteen weeks, although the majority emerged only after about six months. This indicates a much shorter life cycle than is usually estimated for *Hexagenia*. Evidence thus reinforces the conclusion that temperature may determine the length of the life cycle. The number of eggs produced per female varies between 2,200 and 8,000, with the average female producing about 4,000 eggs. The time required for hatching varies depending on the temperature. In the laboratory eggs have been hatched in eleven to twenty-six days, with some eggs hatching much later. In lakes under natural conditions the eggs probably hatch about two weeks after they are laid. In Utah canals the eggs appear to remain dormant during the winter. In the northern part of the range the adults emerge over a period of three to six weeks, with stragglers continuing to appear much later. In Florida *H. munda orlando* emerges principally during the summer; the adults can be taken in nearly every month of the year except in midwinter. Emergence occurs in the late afternoon or at dusk. At the time of emergence the nymphs, which are ordinarily negatively phototropic and positively geotropic, reverse tropisms and swim from the lake or stream bottom to the surface and break through the surface film; the subimagoes immediately burst free. The subimago stage lasts from twenty-four to forty-eight hours.

Fremling (1970) has an excellent account of this genus.

Hexagenia mayflies tend to emerge en masse, and river residents are accustomed to nuisance problems caused by the insects during periods of maximum emergence. Tree limbs droop under their weight, and drifts of the insects form under street lights where they decay and create objectionable odors. Shoppers desert downtown areas as the large, clumsy insects fly in their faces, cover windows, and blanket sidewalks. In extreme cases snowplows are called out to reopen highway bridges which have become impassable. Particles of cast mayfly cuticle cause allergic reactions in people. Mayflies become a hazard to navigation when they are attracted by the powerful arc and mercury-vapor searchlights used by tow-boats to spot unlighted channel markers. Because mayflies cause severe nuisance problems, several river cities have tried unsuccessfully to control them.

The name "Green Bay fly" is often used for the mayfly because people still recall the hordes of *Hexagenia* mayflies which formerly arose from Green Bay of Lake Michigan and literally covered portions of the city of Green Bay, Wisconsin. Because of pollution, Green Bay flies are now rare on the lower reaches of the bay near the mouth of the Fox River (Lee, 1962). Pollution has decimated the *Hexagenia* mayfly population in the western end of Lake Erie (Britt, 1955; Beeton, 1961; Carr and Hiltunen, 1965). *Hexagenia* emergences were once common along the Illinois River, but pollution has virtually eliminated the insects from the upper 150 miles of the river (Richardson, 1921; Mills et al., 1966). *Hexagenia* mayflies, which were once common along the entire Upper Mississippi River, are now rare for 30 miles below Minneapolis, Minnesota, and for almost 200 miles downstream from St. Louis, Missouri (Fremling, 1964). *Hexagenia* and *Pentagenia* mayflies still occur abundantly in the less polluted areas of the Upper Mississippi River. In Pool 19, for example, Carlander et al. (1967) estimated the June, 1962 nymphal *Hexagenia* population to be 23.6 billion.

The average numbers of *H. limbata* nymphs per square foot in a clean Michigan lake ranged from 6 to 30. In Lake Winnipeg estimates of 62,000,000 to 93,700,000 nymphs per square kilometer were given for two species. The estimates were made at the time of maximal population density. In the Michigan lakes which were studied, nymphs comprised as much as 59% of the entire volume of macroscopic bottom fauna in suitable areas.

Adult Characteristics. Wing length: 10–25 mm. Frons does not extend below eyes. Eyes of male separated dorsally by a distance slightly less than width of eye. Crossveins of forewings not crowded at bullae; three to fourteen veins connect vein A_1 to hind margin; wings may be tinged with brown, or pale with or without some crossveins dark margined. Abdomen usually with striking dark pattern. Genital forceps of male three- or four-segmented; penes bladelike. Terminal filament vestigial in both sexes.

Mating Flights. At dusk males of *H. limbata* congregate alongside and above trees on the lakeshore. In Michigan a mating flight was observed along the margin of a lake above vegetation; most of the males flew at treetop level, but some flew thirty to sixty feet above the ground. On joining the swarm the males immediately began the characteristic up-and-down flight; within ten minutes most males had joined the group, each one flying rapidly upward for several feet and then drifting downward to rise again. A short time later the females entered the swarm, flying horizontally through it. Each female was then seized by a male in the typical copulatory position, and the pairs continued to fly but slowly lost altitude. Within a short time, usually less than thirty seconds, the pairs separated; the females left the swarm and flew directly out over the lake, and the males usually returned to the swarm. The males continued to fly until just after dark, then sought resting places where they remained until they died.

To oviposit, some females plunge to the water, and others fly back and forth ten to twenty feet above the water surface for several minutes before dropping to the water. As the females lie on the water with the wings outspread, the last two abdominal segments are raised sharply upward and the eggs are extruded. More rarely, the females barely touch the water surface, release some of their eggs, and then rise again to repeat the performance. Usually within thirty minutes after the first females appear over the lake, all those taking part in the mating flight have oviposited. After ovipositing, the females are generally taken by fish; those not eaten have been observed flying for more than an hour after releasing their eggs.

Taxonomy. Walsh, 1863. *Proc. Entomol. Soc. Philadelphia* 2:197. Type species: *H. limbata* (Serville); type locality: North America, probably the Columbia River area.

Distribution. Probably confined to the western hemisphere. Occurring from the Rio Negro of Argentina north to Great Slave Lake, Northwest Territories, Canada.

Subgenus *Hexagenia* Walsh sensu stricto (figs. 3, 27, 35, 191, 192, 209, 340, 343, 345, 347, 427–429)

Nymphal Characteristics. Not distinguishable from *Pseudeatonica* (see **Distribution**).

Adult Characteristics. Vein A_1 of forewing connected to hind margin by eight to fourteen veins. Genital forceps of male four-segmented. Male genitalia illustrated in fig. 347.

Taxonomy. Spieth (1941b) provided a revision of the species and subspecies of *Hexagenia* and constructed a key to the adults. The keys to the nymphs by Burks (1953) are generally workable. Gooch (1967) has reported on a way to distinguish the two most common species, *H. limbata* and *H. bilineata*, as

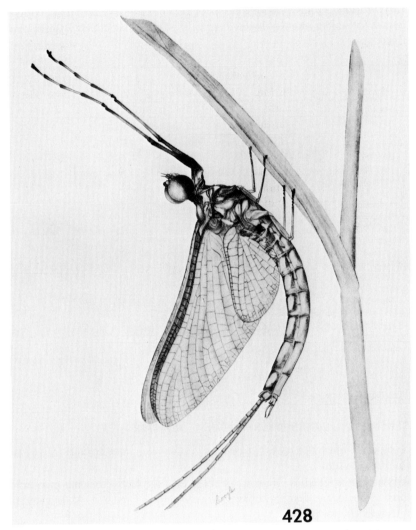

428

Fig. 428. *Hexagenia* (*Hexagenia*) *munda orlando*, male. (From Berner, 1950.)

nymphs, but Fremling (personal communication) reports the character to be unreliable; in mature nymphs, Fremling regards the yellowish coloration of *limbata* and the dark color of *bilineata* to be the most useful criteria for separation of the nymphs. The subspecies problem in this genus appears to be very complex, and it is obvious that additional study is needed. McCafferty (1975b) has reviewed the subgenus *Hexagenia* in the United States.

Distribution. From the Mexican states of Jalisco and Tamaulipas north to British Columbia, the Northwest Territories, Ontario, and New Brunswick of Canada.

Species	United States						Canada			Mexico		Central America
	SE	NE	C	SW	NW	A	E	C	W	N	S	
atrocaudata												
McDunnough	X	X	X	...	...	...	X	...	...	...	...	...
bilineata (Say)	X	...	X	...	...	...	...	...	...	X	...	...
limbata californica												
Upholt	...	...	...	X	...	...	...	...	...	...	...	...
limbata limbata												
Serville	...	...	X	X	X	...	...	X	X	...	...	...
limbata occulta												
Walker	X	X	X	...	...	...	X	X	...	...	...	...
limbata venusta												
Eaton	X	...	X	...	...	...	...	...	...	X	...	...
limbata viridescens												
(Walker)	...	...	X	...	...	...	X	X	...	...	...	...
munda affiliata												
McDunnough	...	X	X	...	...	...	X	...	...	...	...	...
munda elegans												
Traver	X	...	X	...	...	...	...	...	...	...	...	...
munda marilandica												
Traver	X	X	...	...	...	...	...	...	...	...	...	...
munda munda												
Eaton	...	...	X	...	...	...	...	...	...	...	...	...
munda orlando												
Traver	X	...	...	...	...	...	...	...	...	...	...	...
rigida												
McDunnough	X	X	X	...	...	...	X	X	...	...	...	...

Subgenus *Pseudeatonica* Spieth (figs. 341, 348)

Nymphal Characteristics. Not distinguishable from the subgenus *Hexagenia* (see **Distribution**).

Adult Characteristics. Vein A_1 of forewings connected to wing margin by three to six veins. Genital forceps of male three-segmented. Male genitalia illustrated in fig. 348.

Taxonomy. Spieth, 1941b. *Amer. Midland Natur.* 26:269. Type species: *Hexagenia* (*Pseudeatonica*) *mexicana* Eaton. This taxon was described as a subgenus of *Hexagenia* but has been regarded by others variously as a separate genus or as a subgenus of the African genus *Eatonica*. The adults can be identified to species from the key given by Spieth (1941b). McCafferty (1970) has recently discussed the subgeneric status of *Pseudeatonica* and has illustrated the nymph of *H. albivitta* (Walker).

Distribution. From the Rio Negro of Argentina north, presumably into south-

Fig. 429. *Hexagenia (Hexagenia) bilineata* emerging from nymphal skin.

ern Mexico. *H. albivitta* occurs north to Costa Rica. *H. mexicana* occurs in Mexico, according to Eaton (we assume the locality is southern Mexico).

Litobrancha McCafferty (figs. 193, 194, 342, 344, 346)

Nymphal Characteristics. Body length: 20–29 mm. Mandibular tusks with long setae on mesal and lateral margins. Head with frontal process widest near midpoint of length; apex obtusely angulate. Flagella of antennae with only short scattered setae. Gills on segment 1 very small, unbranched. Anterior (dorsal) lamella of gill 2 with outer margin of platelike portion distinctly expanded in basal half. Illustration: Parts in McCafferty, 1971a.

Nymphal Habitat. Small brooks and streams in mixtures of silt and sand in which the nymphs burrow.

Nymphal Habits. The nymphs are negatively phototropic, burrowing into the sand and silt during daylight. Waves of gill motion pass rapidly backward. When a nymph is resting, it holds the gills upright or tilted only slightly to the rear; when it is burrowing, it holds the legs tightly against the head and spread outward as the body lunges forward. The nymphs are good swimmers, using undulating body motions.

Life History. The length of life is not known, but fairly well grown nymphs are common in October. The adults emerge in May or earlier.

Adult Characteristics. Wing length: 15–24 mm. Frons bilobed and extending well below the eyes. Eyes of male separated dorsally by a distance slightly greater than diameter of lateral ocelli. Crossveins of forewings not crowded at bullae; two to four veins connect vein A_1 to wing margin; wings tinged with dark reddish brown. Abdomen yellowish with obscure dark markings. Genital forceps of male four-segmented; penes strongly recurved ventrally at apex (fig. 346). Terminal filament vestigial in both sexes.

Mating Flights. Unreported, except for the fact that the males do not fly in great swarms as in *Hexagenia*.

Taxonomy. McCafferty, 1971a. *J. New York Entomol. Soc.* 79:45. Type species: *L. recurvata* (Morgan). McCafferty isolated the species *L. recurvata* from *Hexagenia* because it is more closely related to *Eatonigenia* (Oriental) and *Eatonica* (Ethiopian) than to the genus *Hexagenia*.

Distribution. Nearctic. The single species is in eastern North America from Michigan, Ontario, and Quebec south to North Carolina.

Subfamily Pentageniinae

Nymphal Characteristics. Frontal process of head bifid. Mandibular tusks with a row of short stout spurs, making the lateral margin appear crenate. Galea-lacinia of maxillae only slightly longer than broad. Maxillary palpi broad, rounded apically, segments 2 and 3 combined less than three times as long as broad.

Adult Characteristics. Pronotum of male reduced, about three times as wide as long. Vein A_1 attached to anal margin by two or three veinlets. Penes of male long and tubular. Caudal filaments of female shorter than body.

Pentagenia Walsh (figs. 30, 41, 211, 214, 430)

Nymphal Characteristics. Body length: 18–24 mm. Mandibular tusks with short setae apically, long setae at base; tusks crenate on outer upper margin. Head with frontal process bifid, the sides divergent apically. Flagella of antennae without setae. Gills on segment 1 bifid to base. Anterior (dorsal) lamella of gill 2 with outer margin of platelike portion expanded only slightly at base. Illustration: Fig. 430.

Nymphal Habitat. Only in large rivers. The nymphs honeycomb hard clay banks with their U-shaped burrows.

Nymphal Habits. Not described.

Life History. The adults have been recorded from the Mississippi River in Iowa as early as mid-June, with continued emergence until early September. In Florida the adults have been taken as early as May 3, in Louisiana as late as October 9. The adults gather at lights quite readily, which accounts for a fair number of records of the genus. Nevertheless, the genus seems to be relatively uncommon throughout its range.

Adult Characteristics. Wing length: 14–19 mm. Frons does not extend below eyes. Eyes of male large and almost contiguous dorsally. Crossveins of forewings not crowded at bullae; two to three veins connect vein A_1 to wing

NOTE: McCafferty and Edmunds have a manuscript in preparation in which the subfamily Pentageniinae is transferred from the family Ephemeridae to the family Palingeniidae.

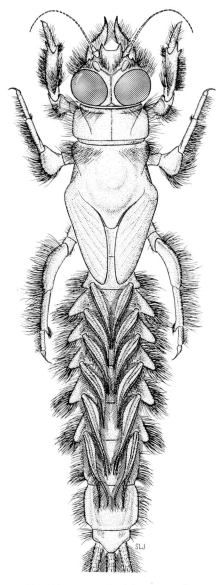

Fig. 430. *Pentagenia vittigera*, nymph.
(From Edmunds, Allen, and Peters, 1963.)

margin; wings pale or with four dark spots at bullae. Abdomen pale with dark medial stripe on terga. Genital forceps of male four-segmented; penes long and tubular. Male genitalia illustrated in fig. 214. Terminal filament somewhat shorter than cerci in female, much shorter than cerci in male.

Mating Flights. Not described.

Taxonomy. Walsh, 1863. *Proc. Entomol. Soc. Philadelphia* 2:196. Type species: *P. vittigera* (Walsh); type locality: Rock Island, Illinois. Of the two species of *Pentagenia*, *P. robusta* McDunnough is known only from the types, but *P. vittigera* is widespread.

Distribution. Nearctic, primarily in central North America from Texas to Florida and north to Manitoba.

Family PALINGENIIDAE

The Palingeniidae are Ephemeroidea that are clearly derived from a proto-*Pentagenia* ancestor. The nymphs of *Pentagenia* have in fact evolved to the palingeniid grade, but the adults have not diverged from the characters of the Ephemeridae. The adults of the Palingeniidae are strongly differentiated from the Ephemeridae, including *Pentagenia*.

The Palingeniidae are widespread in the Old World, where they are found in the large lakes and rivers of Eurasia, New Guinea, Madagascar, and questionably Africa. Except for *Palingenia*, the family is rather poorly known. It is not known to occur in the Americas, but this will change if further studies show that the North American *Pentagenia* should be placed in the Palingeniidae.

Nymphal Characteristics. Mandibular tusks well developed and more or less flattened with the lateral margins serrate to scalloped; the tusks of most Palingeniidae have the outer margins toothed or serrate (in *Pentagenia* spur-like setae make the margin toothed). Legs fossorial. Gills dorsal; gills on segment 1 reduced or absent, gills on segments 2–7 with fringed margins moderately wide.

Adult Characteristics. Wings with numerous crossveins; veins of forewings in geminate pairs, converging at wing margins to various degrees (R_{3b} and IR_3, R_{4+5} and MA_1, IMA and MA_2). Genital forceps with two to six short terminal segments.

NOTE: McCafferty and Edmunds have a manuscript in preparation in which the subfamily Pentageniinae is transferred from the family Ephemeridae to the family Palingeniidae.

Family POLYMITARCYIDAE

The somewhat diverse Polymitarcyidae, whose nymphs are specialized for burrowing, were probably derived from the Euthyplociidae, a family whose nymphs lack specializations for burrowing.

The family is widespread on the continents but absent from the Australasian realm. There are three subfamilies of which two occur in North America. The Asthenopodinae are pantropical and are found in South America, Africa, and the Oriental region. Most species are pale milky white with darker markings, and all seem to have crepuscular to nocturnal mass mating flights.

The family and subfamily names Polymitarcyidae and Polymitarcyinae are reserved for these taxa under rule 40 of the International Code of Zoological Nomenclature. This is essential to avoid the use of two separate names for these taxa; the subjective synonymy of *Polymitarcys* and *Ephoron* is not universally recognized.

Nymphal Characteristics. Maxillary and labial palpi two-segmented; labial palpi held beneath and at right angles to glossae and paraglossae. Mandibular tusks curve downward apically, either the upper surface with numerous tubercles or the mesal surfaces with one or more tubercles. Legs fossorial. Gills dorsal; gills on abdominal segment 1 vestigial. Three caudal filaments.

Adult Characteristics. Wings somewhat translucent, whitish with purplish brown or brownish veins near anterior margin. Veins MP_2 and CuA strongly divergent from MP_1 at base. Middle and hind legs of both sexes and forelegs of female poorly developed or vestigial. Genital forceps of male four-segmented or two-segmented. Two caudal filaments in male, two or three caudal filaments in female.

Subfamily Polymitarcyinae

A single genus, *Ephoron* (*Polymitarcys*) is widespread in the Holarctic, Ethiopian, and Oriental regions. See *Ephoron* for characteristics and for details of Nearctic distribution.

Ephoron Williamson (figs. 28, 34, 36, 38, 40, 199, 203, 205, 431)

Nymphal Characteristics. Body length: 12–17 mm. Mandibular tusks with numerous tubercles on upper and lateral surfaces. Head with rounded median frontal process. Forelegs short and stout, fossorial, tibiae and tarsi not partially fused, cylindrical. Claws of forelegs attached apically. Gills on abdominal segments 1–7, single on segment 1, forked on segments 2–7, each lamella with short fringes, shortest on lateral margins of dorsal lamellae; lateral tracheae numerous. Caudal filaments short. Illustration: Fig. 431.

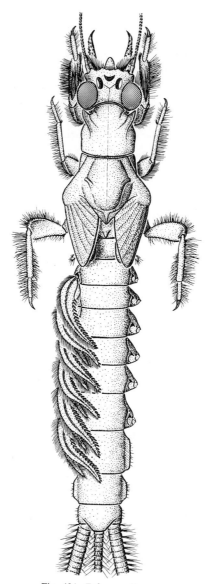

Fig. 431. *Ephoron album*
(right gills removed), nymph.

Nymphal Habitat. Nymphs of *E. leukon* are found in relatively large streams or in gravel and sand shoals in lakes. They are tolerant of a wide variety of currents from lake shoals and slow-moving sections of rivers to the swift riffles of streams. The nymphs live in tubular burrows; during the daylight hours they go fairly deep into the burrows, but as darkness falls they move closer to the surface. It is believed that the majority of the nymphs live in water less than five feet deep.

Ephoron album nymphs occupy a substrate similar to that occupied by *E. leukon* nymphs, but they are also found in sandy clay substrates that are firm and solid. The nymphs also occur in great numbers in clay river banks, where they extensively honeycomb the bank just below the waterline, and in lakes which contain currents produced by wave action.

Nymphal Habits. The nymph creates a current through the U-shaped burrow by almost constant motion of the gills. Small particles suspended in the water are drawn in and caught in the brushes on the forelegs and the head. Nymphs have been observed bringing their forelegs together beneath their mouths; from the movements of the mandibles, it appeared that they were eating debris from the brushes. Nymphs in the laboratory have been observed leaving their burrows at night, raking up slime and debris from the sand, and drawing the material into the burrows. They ate some of the material and swept the remainder out by creating a current with their gills.

Life History. The eggs of *E. album* require a period of cold temperatures to promote hatching. After an interval of dormancy and cold weather they further require temperatures of about 10° C (50° F) for hatching and relatively high temperatures during the growing season of 2½ to 4 months. In all, the eggs require 8 to 9½ months before hatching. The number of instars has not been determined, but it is believed that these nymphs pass through fewer instars than do nymphs of *Ephemera* or *Hexagenia*. Adults have been collected from May to September in various parts of the range. In Lake Erie, where *Ephoron* has been studied intensively, adults were first recorded on July 25, and the last emergence occurred on September 20. In Nevada, Utah, Idaho, and Wyoming emergence occurs primarily in August and September. In emerging, nymphs of *E. album* rise to the surface, immediately shed the nymphal exuviae, and rise from the water as subimagoes. The males require a few minutes at most to molt to the imago stage after emergence. The females do not molt to the imago stage. In Utah the first individuals to emerge are all males, but the females begin to emerge a few minutes later. When nymphs of *E. leukon* are mature, they migrate to shallow water or even onto wet mud at the edge of the water. The subimagoes emerge and take flight at once with the males molting to the adult stage almost immediately. Molting has been reported as occurring in the air during flight, but others have observed males settling on the water or on objects on the shore where they immediately molt.

Adult Characteristics. Wing length: 11–13 mm. Adults milky white with brown to purplish brown shading. Eyes of male small, separated by distance somewhat greater than diameter of eye. Wings of female (subimago) distinctly translucent; wings of male slightly translucent; main veins of costal margin purplish, all others pale. Wings with numerous crossveins and netlike marginal intercalaries. Cubital area with four or more long intercalaries. Hind wing with blunt costal angulation. Forelegs of male about as long as body, femur short; tibia long, 4 to 4½ times length of femur; tarsus slightly shorter than tibia. Male genitalia illustrated in fig. 205; forceps four-segmented. Cerci well developed in male; short rudimentary terminal filament present. Three well-developed caudal filaments in female.

Mating Flights. The flights take place about a foot above the surface of the water and seldom rise more than four or five feet; they do not display the up-and-down movement so characteristic of other mayflies. The males patrol a short stretch of water. The emerging females join the males until mated, then immediately leave the swarm, settle on the water, expel their eggs, and die. Copulation lasts only a few seconds. After copulating, the males return to the swarms. Usually all members of a swarm are dead within an hour and a half after the start of the flight.

Taxonomy. Williamson, 1802. *Trans. Amer. Phil. Soc.* 5:71. Type species: *E. leukon* Williamson; type locality: Passaic River, Belleville, New Jersey. The two species of this genus are readily separable both as adults and as nymphs.

Distribution. Ethiopian, Oriental, Holarctic. In the Nearctic, widespread in southern Canada and the United States south to Georgia, Ohio, Kansas, and New Mexico. *Ephoron leukon* is eastern and *E. album* (Say) is western. Both species occur in Illinois, Indiana, and Ohio.

Subfamily Campsurinae

This subfamily is abundant and diverse in the Neotropical region with progressively less species diversity to the north.

Nymphal Characteristics. Head without a frontal process. Mandibular tusks with one to many tubercles on the mesal margin. Fore tarsi broad and partially fused to tibiae. Claws of forelegs attached subapically.

Adult Characteristics. Forewings with a moderate number of crossveins; few or no marginal intercalaries present. Two long cubital intercalaries present, attached basally to vein CuA. Middle and hind legs reduced to either twisted vestiges or flat blades. Genital forceps of male two-segmented.

Campsurus Eaton (figs. 39, 200, 350, 351, 432)

Nymphal Characteristics. Body length: 10–30 mm. Mandibular tusks with prominent basal or subbasal tubercle on median margin and several to many smaller apical crenations; long setae, usually numerous on lateral margins of mandibles. Illustration: Fig. 432.

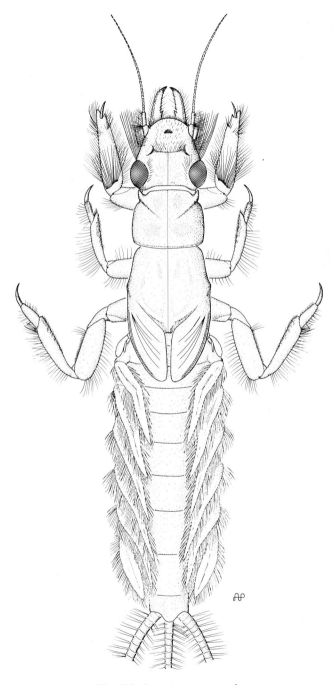

Fig. 432. *Campsurus* sp., nymph.

Nymphal Habitat. Apparently confined to silty mud in a variety of streams. In the northern part of the range the nymphs are found principally in large rivers. The nymphs form U-shaped burrows.

Nymphal Habits. Unknown; probably similar to *Tortopus*.

Life History. Unknown; probably similar to *Tortopus*. The dearth of males in collections suggests that the females may reproduce parthenogenetically, but it is more probable that the males are not as positively phototropic as are the females, which is also the case in the genus *Ephoron*.

Adult Characteristics. Wing length: 5.5–20.0 mm. Middle and hind legs much reduced, broadened, and with femora bladelike; tibiae and tarsi absent. Without clawlike appendage extending from forceps base lateral to forceps. Male genitalia illustrated in fig. 351.

Mating Flights. Described for only a single species, *C. segnis*, from British Guiana. The species was observed flying in large numbers in late evening and in the early hours of darkness. Only two males were observed, although thousands of females gathered at lights. It is likely that the adults mate, drop to the water surface, and die, all in less than an hour. The absence of functional legs suggests that the adult must spend its much abbreviated life on the wing.

Taxonomy. Eaton, 1868. *Entomol. Mon. Mag.* 5:83. Type species: *C. latipennis* (Walker); type locality: Para, Brazil. In the area covered by the present volume there are relatively few species of *Campsurus*. The genus is widely dispersed in South America with many species which differ primarily in the male genitalia. The penes are unusual in being able to revolve in a horizontal plane so that the apex may point either anteriorly, posteriorly, or ventrally. The nymphs cannot be identified to species.

Distribution. Neotropical and Nearctic. In the Nearctic, north to Texas.

Species	United States						Canada			Mexico		Central America
	SE	NE	C	SW	NW	A	E	C	W	N	S	
cuspidatus Eaton	. . .	. . .	. . .	. . .	. . .	. . .	. . .	. . .	. . .	. . .	. . .	X
decoloratus (Hagen) . .	. . .	. . .	X	. . .	. . .	. . .	. . .	. . .	. . .	X	. . .	. . .
emersoni Traver	. . .	. . .	. . .	. . .	. . .	. . .	. . .	. . .	. . .	. . .	. . .	X

Tortopus Needham and Murphy (figs. 33, 206, 207, 349)

Nymphal Characteristics. Body length: 20–35 mm. Mandibular tusks with single prominent subapical tubercle on median margin, although another tubercle may also occur basal to this; long setae usually few on lateral margin of mandibles. Illustration: Scott, Berner, and Hirsch, 1959, figs. 3–14.

Nymphal Habitat. The nymphs burrow in the clay banks of large rivers and in some areas honeycomb the banks at almost every bend. If the banks are composed of suitably firm clay, the nymphs may be located between bends. Occasionally the nymphs burrow into rotten wood; they have also been observed in sandy clay, peat, and unconsolidated sandstone.

Nymphal Habits. The burrows are U-shaped tubes with parallel arms. Apparently the burrow is enlarged in length and diameter as the nymph grows. It is believed that the nymphs feed by filtering the river water moving through the burrow.

Life History. Not known with certainty. The females do not molt to the imago stage. It is believed that one year or less is required for development. In Florida adults emerge from June 6 through November 14, with the beginning and ending dates varying less than one week. Emergence is continuous but rhythmic, with increasing numbers correlated with a decrease in lunar light intensity.

Adult Characteristics. Wing length: 9–19 mm. Middle and hind legs vestigial and twisted, but with all segments present. Male genitalia with clawlike appendage extending laterally from forceps base. Male genitalia illustrated in fig. 206.

Mating Flights. Mass flights occur over rivers from just after sunset to as late as 2:00 A.M.

Taxonomy. Needham and Murphy, 1924. *Bull. Lloyd Lib.* 24, Entomol. Ser., 4:23. Type species: *T. igaranus* Needham and Murphy; type locality: Rio Putumayo between Puerto Alfonso and mouth of Rio Igara-Paraná, Peru (now Colombia). The two species of *Tortopus* whose males are known for the United States and Canada should be easily identifiable from the existing keys and their distribution (see Traver, 1950). The male of *T. unguiculatus* was described from Costa Rica. *T. circumfluus*, from Texas, is known from females only. There are no keys to the nymphs. The nymphs are so similar to those of *Campsurus* that the key characters which separate the two genera should be regarded with caution.

Distribution. Neotropical and Nearctic. In the Nearctic, north to Nebraska, Manitoba, South Carolina, and Florida.

Species	United States						Canada			Mexico		Central America
	SE	NE	C	SW	NW	A	E	C	W	N	S	
circumfluus Ulmer.....	...	...	X	...	...	...	...	...	...	...	...	...
incertus (Traver)	X	...	...	...	...	...	...	...	...	...	...	...
primus (McDunnough)	...	...	X	...	...	...	...	X	...	...	...	...
unguiculatus (Ulmer) ...	...	...	...	...	...	...	...	...	...	...	...	X

APPENDIXES

Appendix I

A Guide to Changes in Scientific Names

The purpose of this guide is to assist the reader to determine the current names of species that appear in other literature under scientific names which have subsequently been changed. Traver (in Needham, Traver, and Hsu, 1935) supplied complete synonymies for North American species and genera, and Burks (1953) updated many synonyms in his discussion of all North American genera and the species in Illinois and surrounding areas. In this guide we have tried to list the correct names of all species that appear in these two works and elsewhere in the literature under names that have since been changed in form or in spelling. (In a few cases we do refer to important synonyms noted by Burks.) We have also attempted to guide workers to changes in the names of Mexican and Central American species. Readers should consult the taxonomic papers for explanations of these changes.

Acentrella, synonym of *Baetis*
Attenuatella, renamed *Attenella*, subgenus of *Ephemerella*
Baetis bundyi, emended to *B. bundyae*
Baetis cingulatus, renamed *Baetis quebecensis*
Baetis herodes, synonym of *B. hageni*
Baetis minimus, synonym of *B. bicaudatus*
Baetis unicolor, renamed *B. hageni*
Baetisca thomsenae, synonym of *B. carolina*
Baetodes sigillatus, synonym of *B. arizonensis*
Blasturus, synonym of *Leptophlebia*; *Blasturus austrinus*, *B. gravastellus*, *B. intermedius*, *B. nebulosus*, and *B. pacificus* now *Leptophlebia austrina*, *L. gravastella*, *L. intermedia*, *L. nebulosa*, and *L. pacifica*
Blasturus vibex, now *Leptophlebia pacifica*
Callibaetis claudiae, name without description, synonym of *C. nigritus*
Callibaetis fuscus, synonym of *C. coloradensis*
Callibaetis hageni, name not applicable to any species with certainty; some of

the specimens assigned to this name belong to *C. carolus*, *C. coloradensis*, and *C. americanus*

Campsurus puella, name not applicable with confidence to any known species (also called *Tortopus puella*)

Chitonophora, synonym of subgenus *Ephemerella*

Cinygmula atlantica, synonym of *C. subaequalis*

Cinygmula hyalina, synonym of *C. par*

Epeorus modestus, now *Stenonema modestum*

Ephemerella angusta, synonym of *E. (Serratella) tibialis*

Ephemerella autumnalis, synonym of *E. (Drunella) spinifera*

Ephemerella bicolorides, synonym of *E. (Eurylophella) verisimilis*

Ephemerella cherokee, synonym of *E. (Drunella) tuberculata*

Ephemerella cognata, synonym of *E. (Serratella) teresa*

Ephemerella columbiella, synonym of *E. (Caudatella) heterocaudata heterocaudata*

Ephemerella consimilis, name not applicable with confidence to any known species

Ephemerella depressa, synonym of *E. (Drunella) cornuta*

Ephemerella euterpe, synonym of *E. (Ephemerella) maculata*

Ephemerella flavitincta, now *E. (Drunella) grandis flavitincta*

Ephemerella glacialis, synonym of *E. (Drunella) grandis ingens*

Ephemerella glacialis carsona, synonym of *E. (Drunella) grandis grandis*

Ephemerella hirsuta, synonym of *E. (Attenella) attenuata*

Ephemerella ingens, now *E. (Drunella) grandis ingens*

Ephemerella lapidula, synonym of *E. (Drunella) flavilinea*

Ephemerella orestes, synonym of *E. (Caudatella) jacobi*

Ephemerella proserpina, synonym of *E. (Drunella) grandis ingens*

Ephemerella sierra, synonym of *E. (Drunella) spinifera*

Ephemerella spinosa, synonym of *E. (Caudatella) hystrix*

Ephemerella unicornis, name not applicable with confidence to any known species

Ephemerella vernalis, synonym of *E. (Ephemerella) rotunda*

Ephemerella wilsoni, synonym of *E. (Drunella) coloradensis*

Ephemerella yosemite, synonym of *E. (Drunella) grandis ingens*

Habrophlebia pusilla, synonym of *H. vibrans*

Heptagenia rubroventris, synonym of *H. rosea*

Hexagenia recurvata, now in *Litobrancha*

Iron, now subgenus of *Epeorus*

Iron confusus, synonym of *Epeorus (Iron) pleuralis*

Iron fraudator, synonym of *Epeorus (Iron) pleuralis*

Iron humeralis, synonym of *Epeorus (Iron) vitreus*

Iron modestus, now *Stenonema modestum*

Iron proprius, synonym of *Epeorus (Iron) longimanus*

Iron sancta-gabriel, synonym of *Epeorus (Iron) hesperus*

Iron tenuis, synonym of *Epeorus (Iron) fragilis*

Iron youngi, synonym of *Epeorus (Iron) albertae*

Ironopsis, subgenus of *Epeorus*

Isonychia campestris and *I. manca*, now subspecies of *I. sicca*

Leptohyphes mithras, now in *Haplohyphes*

Leptophlebia vibex, synonym of *L. pacifica*

Metretopus norvegicus, synonym of *M. borealis*

Metreturus, synonym of *Acanthametropus*

Neobaetis, synonym of *Callibaetis*

Neobaetis paulinus, now in *Callibaetis*

Neocloeon, synonym of *Cloeon*

Oligoneuria ammophila, now in *Homoeoneuria*

Oreianthus, synonym of *Neoephemera*

Paraleptophlebia compar, synonym of *P. associata*

Paraleptophlebia invalida, synonym of *P. gregalis*

Paraleptophlebia pallipes, renamed *P. memorialis*

Paraleptophlebia sculleni, synonym of *P. gregalis*

Pentageniidae, subfamily Pentageniinae of Ephemeridae (but see footnote in
    discussion of family Palingeniidae)

Pseudeatonica, subgenus of *Hexagenia*

Pseudocloeon minutum, now *Paracloeodes minutus*

Rheobaetis, synonym of *Heterocloeon*

Rheobaetis traverae, synonym of *Heterocloeon curiosum*

Rhithrogena brunnea, renamed *R. hageni*

Rhithrogena doddsi, synonym of *R. hageni*

Rhithrogena petulans, synonym of *R. morrisoni*

Siphlonurus inflatus, synonym of *S. occidentalis*

Siphlonurus maria, synonym of *S. spectabilis*

Stenacron areion, synonym of *S. interpunctatum canadense*

Stenacron canadense, now *S. interpunctatum canadense*

Stenacron frontale, now *S. interpunctatum frontale*

Stenacron heterotarsale, now *S. interpunctatum heterotarsale*

Stenacron interpunctatum, now *S. interpunctatum interpunctatum*

Stenonema: S. canadense, S. candidum, S. carolina, S. floridense, S. fron-
    tale, S. gildersleevei, S. heterotarsale, S. interpunctatum, S. min-
    netonka, and *S. pallidum* now in *Stenacron*

Stenonema: S. interpunctatum canadense, S. i. frontale, S. i. heterotarsale,
    and *S. i. interpunctatum* now in *Stenacron*

Stenonema affine, synonym of *Stenacron interpunctatum interpunctatum*

Stenonema alabamae, synonym of *S. exiguum*

Stenonema areion, synonym of *Stenacron interpunctatum canadense*

Stenonema birdi, synonym of *S. tripunctatum tripunctatum*

Stenonema conjunctum, synonym of *Stenacron interpunctatum interpunctatum*

Stenonema metriotes, now *S. integrum integrum*

Stenonema proximum, synonym of *Stenacron interpunctatum frontale*

Stenonema rivulicolum, now *S. fuscum rivulicolum*

Stenonema scitulum, now *S. tripunctatum scitulum*

Stenonema varium, synonym of *S. rubrum*

Stenonema wabasha, now *S. integrum wabasha*

Thraulus albertanus, *T. ehrhardti*, *T. presidianus*, and *T. primanus* now *Traverella albertana*, *T. ehrhardti*, *T. presidiana*, and *T. primana*

Tortopus puella, name not applicable with confidence to any known species, (also called *Campsurus puella*)

Tricorythodes fallacina and *T. fallax*, synonyms of *T. minutus*

Appendix II

Suppliers of Collecting and Rearing Equipment

BioQuip Products, Inc.
P.O. Box 61
El Segundo, California 90406
(Aquatic and terrestrial nets, Tropics nets,
light traps, Malaise traps, etc.)

C. James Brindell
P.O. Box 10015
110 E. Park Avenue
Tallahassee, Florida 32302
(Subimago cages.)

Carolina Biological Supply Company
Box 7
Gladstone, Oregon 97027

2700 York Road
Burlington, North Carolina 27215

(Aquatic and terrestrial nets, Peterson dredges, etc.)

Entomological Supplies, Inc.
5655 Oregon Avenue
Baltimore, Maryland 21227
(Peterson dredges, artificial substrate samplers,
portable aerators, etc.)

D. A. Focks and Associates
P.O. Box 12852, University Station
Gainesville, Florida 32604
(Malaise traps.)

Turtox/Cambosco, Macmillan Science Company, Inc.
8200 South Hoyne Avenue
Chicago, Illinois 60620
(Aquatic and terrestrial nets, scraper nets,
dredge nets, hand screens, artificial substrate samplers, etc.)

Ward's Natural Science Establishment, Inc.
P.O. Box 1712
Rochester, New York 14603

P.O. Box 1749
Monterey, California 93940

(Aquatic and terrestrial nets, drift nets, apron
nets, Ekman and Peterson dredges, artificial substrate
samplers, etc.)

Wildco Instruments
301 Cass Street
Saginaw, Michigan 48602
(Dredge nets, Ekman and Peterson dredges, artificial
substrate samplers, etc.)

BIBLIOGRAPHY

Bibliography

The authors have attempted to supply bibliographic information on all publications from the period 1935–74 which describe new species or establish name changes for the mayfly genera of North America north of Mexico and to provide a complete taxonomic bibliography for Mexico and Central America. In addition, a number of important publications which appear before 1935 and a few papers published since 1974 are listed. Complete bibliographies of the literature published before 1935 can be found in Needham, Traver, and Hsu (1935) or Burks (1953). References to the literature published since 1974 can be found in *Eatonia*, a newsletter for ephemeropterists, edited by Janice G. Peters and published by the S. H. Coleman Library, Florida Agricultural and Mechanical University, Tallahassee, Florida.

Allen, R. K. 1959. A new species of *Ephemerella* from Oregon (Ephemeroptera: Ephemerellidae). *J. Kans. Entomol. Soc.* 32:59–60.
_____. 1965. The adult stages of *Ephemerella* (*Drunella*) *pelosa* Mayo (Ephemeroptera: Ephemerellidae). *Pan-Pac. Entomol.* 41:280–82.
_____. 1966a. New species of *Heptagenia* from western North America (Ephemeroptera: Heptageniidae). *Can. Entomol.* 98:80–82.
_____. 1966b. *Haplohyphes*, a new genus of Leptohyphinae (Ephemeroptera: Tricorythidae). *J. Kans. Entomol. Soc.* 39:565–68.
_____. 1967. New species of New World Leptohyphinae (Ephemeroptera: Tricorythidae). *Can. Entomol.* 99:350–75.
_____. 1968. New species and records of *Ephemerella* (*Ephemerella*) in western North America (Ephemeroptera: Ephemerellidae). *J. Kans. Entomol. Soc.* 41:557–67.
_____. 1973. Generic revisions of mayfly nymphs. I. *Traverella* in North and Central America (Leptophlebiidae). *Ann. Entomol. Soc. Amer.* 66:1287–95.
_____. 1974. *Neochoroterpes*, a new subgenus of *Choroterpes* Eaton from North America (Ephemeroptera: Leptophlebiidae). *Can. Entomol.* 106:161–68.
Allen, R. K., and R. C. Brusca. 1973. New species of Leptohyphinae from Mexico and Central America (Ephemeroptera: Tricorythidae). *Can. Entomol.* 105:83–95.
Allen, R. K., and E. S. M. Chao. 1972. A new species of *Baetodes* from Arizona (Ephemeroptera: Baetidae). *Bull. S. Calif. Acad. Sci.* 71:52.
Allen, R. K., and D. L. Collins. 1968. A new species of *Ephemerella* (*Serratella*) from California (Ephemeroptera: Ephemerellidae). *Pan-Pac. Entomol.* 44:122–24.

314 Bibliography

Allen, R. K., and G. F. Edmunds, Jr. 1958. A new species of *Ephemerella* from Georgia (Ephemeroptera: Ephemerellidae). *J. Kans. Entomol. Soc.* 31:222–24.

———. 1959. A revision of the genus *Ephemerella* (Ephemeroptera: Ephemerellidae). I. The subgenus *Timpanoga*. *Can. Entomol.* 91:51–58.

———. 1961a. A revision of the genus *Ephemerella* (Ephemeroptera: Ephemerellidae). II. The subgenus *Caudatella*. *Ann. Entomol. Soc. Amer.* 54:603–12.

———. 1961b. A revision of the genus *Ephemerella* (Ephemeroptera: Ephemerellidae). III. The subgenus *Attenuatella*. *J. Kans. Entomol. Soc.* 34:161–73.

———. 1962a. A revision of the genus *Ephemerella* (Ephemeroptera: Ephemerellidae). IV. The subgenus *Dannella*. *J. Kans. Entomol. Soc.* 35:333–38.

———. 1962b. A revision of the genus *Ephemerella* (Ephemeroptera: Ephemerellidae). V. The subgenus *Drunella* in North America. *Misc. Publ. Entomol. Soc. Amer.* 3:147–79.

———. 1963a. A revision of the genus *Ephemerella* (Ephemeroptera: Ephemerellidae). VI. The subgenus *Serratella* in North America. *Ann. Entomol. Soc. Amer.* 56:583–600.

———. 1963b. A revision of the genus *Ephemerella* (Ephemeroptera: Ephemerellidae). VII. The subgenus *Eurylophella*. *Can. Entomol.* 95:597–623.

———. 1965. A revision of the genus *Ephemerella* (Ephemeroptera: Ephemerellidae). VIII. The subgenus *Ephemerella* in North America. *Misc. Publ. Entomol. Soc. Amer.* 4:243–82.

———. 1968. A new synonymy in *Ephemerella*. *Ann. Entomol. Soc. Amer.* 61:1044.

Allen, R. K., and S. S. Roback. 1969. New species and records of New World Leptohyphinae (Ephemeroptera: Tricorythidae). *J. Kans. Entomol. Soc.* 42:372–79.

Argyle, D. W., and G. F. Edmunds, Jr. 1962. Mayflies (Ephemeroptera) of the Curecanti Reservoir Basins. Gunnison River, Colorado. *Univ. Utah Anthropol. Pap.* 59:179–89.

Bartlett, L. M. 1941. *Iron fraudator* Traver vs. *Iron pleuralis* Banks (Ephemerida). *Can. Entomol.* 73:218–19.

Beeton, Alfred M. 1961. Environmental changes in Lake Erie. *Trans. Amer. Fish. Soc.* 90:153–59.

Berner, L. 1940a. *Baetisca rogersi*, a new mayfly from northern Florida. *Can. Entomol.* 72:156–60.

———. 1940b. Baetine mayflies from Florida (Ephemeroptera). *Fla. Entomol.* 23:33–45, 49–62.

———. 1946. New species of Florida mayflies (Ephemeroptera). *Fla. Entomol.* 28:60–82.

———. 1948. A new species of mayfly from Tennessee. *Entomol. News* 59:117–20.

———. 1950. The mayflies of Florida. *Univ. Fla. Studies, Biol. Sci. Ser.*, no. 4.

———. 1953. New mayfly records from Florida and a description of a new species. *Fla. Entomol.* 36:145–49.

———. 1955a. The southeastern species of Baetisca (Ephemeroptera: Baetiscidae). *Quart. J. Fla. Acad. Sci.* 18:1–19.

———. 1955b. A new species of Paraleptophlebia from the Southeast (Ephemeroptera: Leptophlebiidae). *Proc. Entomol. Soc. Wash.* 57:245–47.

———. 1956. The genus *Neoephemera* in North America (Ephemeroptera: Neoephemeridae). *Ann. Entomol. Soc. Amer.* 49:33–42.

———. 1958. A list of mayflies from the lower Apalachicola River drainage. *Quart. J. Fla. Acad. Sci.* 21:25–31.

———. 1959a. A tabular summary of the biology of North American mayfly nymphs (Ephemeroptera). *Bull. Fla. State Mus., Biol. Sci. Ser.*, 4:1–58.

———. 1959b. Newfoundland mayflies (Ephemeroptera). *Opuscula Entomol.* 24:212–14.

———. 1975. The mayfly family Leptophlebiidae in the southeastern United States. *Fla. Entomol.* 58(3):137–56.

Berner, L., and R. K. Allen. 1961. Southeastern species of the mayfly subgenus *Serratella* (Ephemerella: Ephemerellidae). *Fla. Entomol.* 44:149–58.

Berner, L., and T. B. Thew. 1961. Comments on the mayfly genus *Campylocia* with a description of a new species (Euthyplociidae: Euthyplociinae). *Amer. Midland Natur.* 66:329–36.

Brekke, R. 1938. The Norwegian mayflies (Ephemeroptera). *Nor. Entomol. Tidsskr.* 5:55–73.

Brinck, P. 1957. Reproductive system and mating in Ephemeroptera. *Opuscula Entomol.* 22:1–37.

Britt, N. W. 1955. *Hexagenia* (Ephemeroptera) population recovery in western Lake Erie following the 1953 catastrophe. *Ecology* 36:520–22.

Britt, N. W., J. T. Addis, and R. Engel. 1973. Limnological studies of the island area of Western Lake Erie. *Bull. Ohio Biol. Surv.* (new series) 4(3).

Brusca, R. C. 1971. A new species of *Leptohyphes* from Mexico (Ephemeroptera: Tricorythidae). *Pan-Pac. Entomol.* 47:146–48.

Brusca, R. C., and R. K. Allen. 1973. A new species of *Choroterpes* from Mexico (Ephemeroptera: Leptophlebiidae). *J. Kans. Entomol. Soc.* 46:137–39.

Burks, B. D. 1947. New Heptagenine mayflies. *Ann. Entomol. Soc. Amer.* 39:607–15.

_____. 1947. New species of *Ephemerella* from Illinois (Ephemeroptera). *Can. Entomol.* 79:232–36.

_____. 1953. The mayflies or Ephemeroptera of Illinois. *Ill. Natur. Hist. Surv. Bull.* 26(1).

Carlander, K. D., C. A. Carlson, V. Gooch, and T. L. Wenko. 1967. Populations of *Hexagenia* mayfly naiads in Pool 19, Mississippi River, 1959–1963. *Ecology* 48:873–78.

Carr, J. F., and J. K. Hiltunen. 1965. Changes in the bottom fauna of western Lake Erie from 1930 to 1961. *Limnol. Oceanogr.* 10:551–69.

Clifford, H. F. 1969. Limnological features of a northern brown-water stream, with special reference to the life histories of the aquatic insects. *Amer. Midland Natur.* 82:578–97.

_____. 1970. Analysis of a northern mayfly (Ephemeroptera) population with special reference to allometry of size. *Can. J. Zool.* 50:957–83.

Clifford, H. F., M. R. Robertson, and K. A. Zelt. 1973. Life cycle patterns of mayflies (Ephemeroptera) from some streams of Alberta, Canada. In *Proceedings of the First International Conference on Ephemeroptera*, W. L. Peters and J. G. Peters, eds., pp. 122–31. E. J. Brill, Leiden.

Coffman, W. P., K. W. Cummins, and J. C. Wuycheck. 1971. Energy flow in a woodland stream ecosystem. I. Tissue support trophic structure of the autumnal community. *Arch. Hydrobiol.* 68:232–76.

Cohen, S. D., and R. K. Allen. 1972. New species of *Baetodes* from Mexico and Central America. *Pan-Pac. Entomol.* 48:123–35.

Coleman, M. J., and H. B. N. Hynes. 1970. The life histories of some Plecoptera and Ephemeroptera in a southern Ontario stream. *Can. J. Zool.* 48:1333–39.

Crass, R. S. 1947. The may-flies (Ephemeroptera) of Natal and the Eastern Cape. *Ann. Natal Mus.* 11:37–110.

Cummins, K. W. 1973. Trophic relations of aquatic insects. *Annu. Rev. Entomol.* 18:183–206.

Daggy, R. H. 1945. New species and previously undescribed naiads of some Minnesota mayflies (Ephemeroptera). *Ann. Entomol. Soc. Amer.* 38:373–96.

Day, W. C. 1952. New species and notes on California mayflies (Ephemeroptera). *Pan-Pac. Entomol.* 28:17–39.

_____. 1953. A new mayfly genus from California (Ephemeroptera). *Pan-Pac. Entomol.* 29:19–24.

_____. 1954a. New species and notes on California mayflies. II. *Pan-Pac. Entomol.* 30:15–29.

_____. 1954b. New species of California mayflies in the genus *Baetis*. *Pan-Pac. Entomol.* 30:29–34.

_____. 1955. New genera of mayflies from California (Ephemeroptera). *Pan-Pac. Entomol.* 31:121–37.

_____. 1956. Ephemeroptera. In *Aquatic insects of California*, R. L. Usinger, ed., pp. 79–105. Univ. Calif. Press, Berkeley.

_____. 1957. The California mayflies of the genus *Rhithrogena*. *Pan-Pac. Entomol.* 33:1–7.

Demoulin, G. 1955. Une mission biologique Belge au Brésil. Ephéméroptères. *Bull. Inst. Roy. Sci. Natur. Belg.* 31(20):1–32.

_____. 1966. Contribution à l'étude des Ephéméroptères de Surinam. *Bull. Inst. Roy. Sci. Natur. Belg.* 42(37):1–22.

Eaton, A. E. 1871. A monograph on the Ephemeridae. *Trans. Entomol. Soc. London*, 1871:1–164.

———. 1883–88. A revisional monograph of recent Ephemeridae or mayflies. *Trans. Linn. Soc. London*, 2d Ser. Zool., No. 3.

———. 1892. *Biologia Centrali-Americana: Insecta, Neuroptera, Ephemeridae*. Vol. 38. Bernard Quaritch, Ltd., London.

Edmunds, G. F., Jr. 1948a. The nymph of *Ephoron album* (Ephemeroptera). *Entomol. News* 59:12–14.

———. 1948b. A new genus of mayflies from western North America (Leptophlebiinae). *Proc. Biol. Soc. Wash.* 61:141–48.

———. 1951. New species of Utah mayflies. I. Oligoneuriidae (Ephemeroptera). *Proc. Entomol. Soc. Wash.* 53:327–31.

———. 1954a. New species of Utah mayflies. II. Baetidae, *Centroptilum* (Ephemeroptera). *Proc. Entomol. Soc. Wash.* 56:1–4.

———. 1954b. The mayflies of Utah. *Proc. Utah Acad. Sci. Arts Lett.* 31:64–66.

———. 1955. Ephemeroptera. In *Systematic entomology: A century of progress in the natural sciences*, pp. 509–12. Calif. Acad. Sci., San Francisco.

———. 1957. *Metretopus borealis* (Eaton) in Canada (Ephemeroptera; Ametropodidae). *Can. J. Zool.* 35:161–62.

———. 1959a. Ephemeroptera. In *Freshwater Biology* (2d ed.), W. T. Edmondson et al., pp. 908–16. Wiley, New York.

———. 1959b. The mayflies of the Glen Canyon Dam area, Colorado River, Utah. *Proc. Utah Acad. Sci. Arts Lett.* 36:79–80.

———. 1959c. Subgeneric groups within the mayfly genus *Ephemerella* (Ephemeroptera: Ephemerellidae). *Ann. Entomol. Soc. Amer.* 52:543–47.

———. 1960. The mayfly genus *Baetisca* in western North America. (Ephemeroptera: Baetiscidae). *Pan-Pac. Entomol.* 36:102–4.

———. 1961a. A key to the genera of known nymphs of the Oligoneuriidae (Ephemeroptera). *Proc. Entomol. Soc. Wash.* 63:255–56.

———. 1961b. Ephemeroptera. In *Encyclopedia of biological sciences*, Peter Gray, ed., pp. 359–61. Reinhold, New York.

———. 1962a. The principles applied in determining the hierarchic level of the higher categories of Ephemeroptera. *Syst. Zool.* 11:22–31.

———. 1962b. The type localities of the Ephemeroptera of North America north of Mexico. *Univ. Utah Biol. Ser.*, No. 12.

———. 1965. The classification of Ephemeroptera in relation to the evolutionary grade of nymphal and adult stages. *Proc. XIIth International Congress of Entomology*, p. 112.

———. 1971. A new name for a subgeneric homonym in *Ephemerella* (Ephemeroptera: Ephemerellidae). *Proc. Entomol. Soc. Wash.* 73:152.

———. 1972. Biogeography and evolution of Ephemeroptera. *Annu. Rev. Entomol.* 17:21–42.

———. 1973. Some critical problems of family relationships in the Ephemeroptera. In *Proceedings of the First International Conference on Ephemeroptera*, W. L. Peters and J. G. Peters, eds., pp. 145–54. E. J. Brill, Leiden.

———. 1974. Some taxonomic changes in Baetidae. *Proc. Entomol. Soc. Wash.* 76:289.

Edmunds. G. F., Jr., and R. K. Allen. 1957a. A checklist of the Ephemeroptera of North America north of Mexico. *Ann. Entomol. Soc. Amer.* 50:317–24.

———. 1957b. A new species of *Baetis* from Oregon (Ephemeroptera: Baetidae). *J. Kans. Entomol. Soc.* 30:57–58.

———. 1964. The Rocky Mountain species of *Epeorus* (*Iron*) Eaton (Ephemeroptera: Heptageniidae). *J. Kans. Entomol. Soc.* 37:275–88.

———. 1966. The significance of nymphal stages in the study of Ephemeroptera. *Ann. Entomol. Soc. Amer.* 59:300–3.

Edmunds, G. F., Jr., R. K. Allen, and W. L. Peters. 1963. An annotated key to the nymphs of the families and subfamilies of mayflies (Ephemeroptera). *Univ. Utah Biol. Ser.*, No. 13.

Edmunds, G. F., Jr., L. Berner, and J. R. Traver. 1958. North American mayflies of the family Oligoneuriidae. *Ann. Entomol. Soc. Amer.* 51:375–82.

Edmunds. G. F., Jr., and S. L. Jensen. 1974. A new genus and subfamily of North American Heptageniidae (Ephemeroptera). *Proc. Entomol. Soc. Wash.* 76:495–97.

Edmunds, G. F., Jr., and R. W. Koss. 1972. A review of the Acanthametropodinae with a description of a new genus (Ephemeroptera: Siphlonuridae). *Pan-Pac. Entomol.* 48:136–44.

Edmunds, G. F., Jr., and G. G. Musser. 1960. The mayfly fauna of Green River in the Flaming Gorge Reservoir Basin of Wyoming and Utah. *Univ. Utah Anthropol. Pap.* 48:111–23.

Edmunds, G. F., Jr., and J. R. Traver. 1954. An outline of reclassification of the Ephemeroptera. *Proc. Entomol. Soc. Wash.* 56:236–40.

———. 1959. The classification of the Ephemeroptera. I. Ephemeroidea: Behningiidae. *Ann. Entomol. Soc. Amer.* 52:43–51.

Flowers, R. W., and W. L. Hilsenhoff. 1975. Heptageniidae (Ephemeroptera) of Wisconsin. *Great Lakes Entomol.* 8:201–18.

Fremling, C. R. 1960. Biology of a large mayfly, *Hexagenia bilineata* (Say) of the Upper Mississippi River. *Iowa Agr. Home Econ. Exp. Sta. Res. Bull.* 482:842–52.

———. 1964. Rhythmic *Hexagenia* mayfly emergences and the environmental factors which influence them. *Verh. Internat. Verein. Limnol.* 15:912–16.

———. 1970. *Mayfly distribution as a water quality index.* United States Environmental Protection Agency, Water Pollution Control Research Series, Report No. 16030 DQH 11/70.

———. 1973. Factors influencing the distribution of burrowing mayflies along the Mississippi River. In *Proceedings of the First International Conference on Ephemeroptera,* W. L. Peters and J. G. Peters, eds., pp. 12–25. E. J. Brill, Leiden.

Gilpin, B. R., and M. A. Brusven. 1970. Food habits and ecology of mayflies of the St. Maries River in Idaho. *Melanderia* (Wash. State Entomol. Soc.) 4:19–40.

Gooch, V. 1967. Identification of *Hexagenia bilineata* and *H. limbata* nymphs. *Entomol. News* 78:101–3.

Hagen, H. A. 1861. Synopsis of the Neuroptera of North America, with a list of South American species: Ephemeridae. *Smithsonian Misc. Coll.* 4:38–55.

———. 1868. On *Lachlania abnormis*, a new genus and species from Cuba belonging to the Ephemerina. *Proc. Bost. Soc. Nat. Hist.* 11:372–74.

Hall, R. J., L. Berner, and E. F. Cook. 1975. Observations on the biology of *Tricorythodes atratus* McDunnough (Ephemeroptera: Tricorythidae). *Proc. Entomol. Soc. Wash.* 77:34–49.

Harper, F. 1970. A new species of *Ameletus* (Ephemeroptera, Siphlonuridae) from Southern Ontario. *Can. J. Zool.* 48:603–4.

Harper, F., and E. Magnin. 1971. Emergence saisonnière de quelques éphéméroptères d'un ruisseau des Laurentides. *Can. J. Zool.* 49:1209–21.

Hashagen, K. A., Jr. 1969. The occurrence of *Hexagenia limbata californica* (Upholt) in a California warmwater reservoir (Ephemeroptera: Ephemeridae). *Wasmann J. Biol.* 27:327–28.

Hilsenhoff, W. L. 1970. *Key to genera of Wisconsin Plecoptera (stonefly) nymphs, Ephemeroptera (mayfly) nymphs, Trichoptera (caddisfly) larvae.* Wisconsin Department of Natural Resources, Research Report 67.

———. 1975. *Aquatic insects of Wisconsin with generic keys and notes on biology, ecology, and distribution.* Wisconsin Department of Natural Resources, Technical Bulletin 89.

Hynes, H. B. N. 1970. *The Ecology of Running Waters.* Univ. Toronto Press, Toronto.

Ide, F. P. 1937. Descriptions of eastern North American species of Baetine mayflies with particular reference to the nymphal stages. *Can. Entomol.* 69:219–31, 235–43.

———. 1941. Mayflies of two tropical genera, *Lachlania* and *Campsurus*, from Canada with descriptions. *Can. Entomol.* 73:153–56.

———. 1954. The nymph of *Rhithrogena impersonata* (Ephemerida) and a new closely related species from the same locality in southern Ontario. *Can. Entomol.* 86:348–56.

———. 1955. Two species of mayflies representing southern groups occurring at Winnipeg, Manitoba (Ephemeroptera). *Ann. Entomol. Soc. Amer.* 48:15–16.

Jensen, S. L. 1969. A new species of *Pseudocloeon* from Idaho (Ephemeroptera: Baetidae). *Pan-Pac. Entomol.* 45:14–15.

———. 1974. A new genus of mayflies from North America (Ephemeroptera: Heptageniidae). *Proc. Entomol. Soc. Wash.* 76:225–28.

Jensen, S. L., and G. F. Edmunds, Jr. 1966. A new species of *Ephemerella* from western North America (Ephemeroptera: Ephemerellidae). *J. Kans. Entomol. Soc.* 39:576–79.

Kapur, A. P., and M. B. Kripalani. 1963. The mayflies (Ephemeroptera) from the northwestern Himalaya. *Rec. Indian Mus.*, 59:183–221.

Keffermüller, M. 1972. Investigations on the fauna Ephemeroptera in Wielkopolska (Great Poland). IV. Analysis of *Baetis tricolor* Tsher. variability and a description of *B. calcaratus* sp. n. *Poznan Towarz. Przy. Nauk* (Biol.) 35:199–226.

Kilgore, J. I., and R. K. Allen. 1973. Mayflies of the Southwest: New species, descriptions, and records (Ephemeroptera). *Ann. Entomol. Soc. Amer.* 66:321–32.

Kimmins, D. E. 1934. Notes on the Ephemeroptera of the Godman and Salvin collection, with descriptions of two new species. *Ann. Mag. Natur. Hist.*, ser. 10, 14:338–53.

———. 1960a. The Ephemeroptera types of species described by A. E. Eaton, R. McLachlan, and F. Walker. *Bull. Brit. Mus. (Natur. Hist.) Entomol.* 9:269–318.

———. 1960b. Notes on East African Ephemeroptera, with descriptions of new species. *Bull. Brit. Mus. (Natur. Hist.) Entomol.* 9:337–55.

Koss. R. W. 1966. A new species of *Thraulodes* from New Mexico (Ephemeroptera: Leptophlebiidae) *Mich. Entomol.* 1:91–94.

———. 1968. Morphology and taxonomic use of Ephemeroptera eggs. *Ann. Entomol. Soc. Amer.* 61:696–721.

———. 1972. Baetodes: New species and new records for North America (Ephemeroptera: Baetidae). *Entomol. News* 83:93–102.

Koss, R. W., and G. F. Edmunds, Jr. 1970. A new species of *Lachlania* from New Mexico with notes on the genus (Ephemeroptera: Oligoneuriidae). *Proc. Entomol. Soc. Wash.* 72:55–65.

Landa, V. 1959. Problems of internal anatomy of Ephemeroptera and their relation to the phylogeny and systematics of their order. *XVth Intern. Cong. Zool. Pap.* 54:1–2.

———. 1968. Developmental cycles of Central European Ephemeroptera and their interrelationships. *Acta Entomol. Bohemoslov.* 65:276–84.

Lee, J. 1962. Sign of a sick river: Green Bay flies are gone. *Green Bay Press-Gazette*, Feb. 25, p. M-6.

Lehmkuhl, D. M. 1968. Observations on the life histories of four species of *Epeorus* in western Oregon (Ephemeroptera: Heptageniidae). *Pan-Pac. Entomol.* 44:129–37.

———. 1970. Mayflies in the South Saskatchewan River; pollution indicators. *Blue Jay* 28:183–86.

———. 1972. *Baetisca* (Ephemeroptera: Baetiscidae) from the western interior of Canada with notes on the life cycle. *Can. J. Zool.* 50:1015–17.

———. 1973a. A new species of *Baetis* (Ephemeroptera) from ponds in the Canadian Arctic, with biological notes. *Can. Entomol.* 105:343–46.

———. 1973b. Adaptations to a marine climate illustrated by Ephemeroptera. In *Proceedings of the First International Conference on Ephemeroptera*, W. L. Peters and J. G. Peters, eds., pp. 33–38. E. J. Brill, Leiden.

Lehmkuhl, D. M., and N. H. Anderson. 1970. Observations on the biology of *Cinygmula reticulata* McDunnough in Oregon (Ephemeroptera: Heptageniidae). *Pan-Pac. Entomol.* 46:268–74.

———. 1971. Contributions to the biology and taxonomy of the *Paraleptophlebia* of Oregon (Ephemeroptera: Leptophlebiidae). *Pan-Pac. Entomol.* 47:85–93.

Leonard, J. W. 1949. The nymph of *Ephemerella excrucians* Walsh. *Can. Entomol.* 81:158–59.

———. 1950. A new *Baetis* from Michigan (Ephemeroptera). *Ann. Entomol. Soc. Amer.* 43:155–59.

Leonard, J. W., and F. A. Leonard. 1962. Mayflies of Michigan trout streams. *Cranbrook Inst. Sci. Bull.*, No. 43.

Lewis, P. A. 1973. Description and ecology of three *Stenonema* mayfly nymphs. In *Proceedings of the First International Conference on Ephemeroptera*, W. L. Peters and J. G. Peters, eds., pp. 64–72. E. J. Brill, Leiden.

_____. 1974a. *Taxonomy and ecology of Stenonema mayflies (Heptageniidae: Ephemeroptera)*. United States Environmental Protection Agency, Environmental Monitoring Series, Report No. EPA-670/4-74-006, pp. 1–81.

_____. 1974b. Three new *Stenonema* species from eastern North America (Heptageniidae: Ephemeroptera). *Proc. Entomol. Soc. Wash.* 76:347–55.

Lyman, F. E. 1944. Taxonomic notes on *Brachycercus lacustris* (Needham) (Ephemeroptera). *Entomol. News* 55:3–4.

McCafferty, W. P. 1968. The mayfly genus *Hexagenia* in Mexico (Ephemeroptera: Ephemeridae). *Proc. Entomol. Soc. Wash.* 70:358–59.

_____. 1970. Neotropical nymphs of the genus *Hexagenia* (Ephemeroptera: Ephemeridae). *J. Georgia Entomol. Soc.* 5:224–28.

_____. 1971a. New genus of mayflies from Eastern North America (Ephemeroptera: Ephemeridae). *J. New York Entomol. Soc.* 79:45–51.

_____. 1971b. New burrowing mayflies from Africa (Ephemeroptera: Ephemeridae). *J. Entomol. Soc. S. Afr.* 34:57–62.

_____. 1972. Pentageniidae, a new family of Ephemeroidea (Ephemeroptera). *J. Georgia Entomol. Soc.* 7:51–56.

_____. 1975a. A new synonym in *Potamanthus* (Ephemeroptera: Potamanthidae). *Proc. Entomol. Soc. Wash.* 77:224.

_____. 1975b. The burrowing mayflies (Ephemeroptera: Ephemeroidea) of the United States. *Trans. Amer. Entomol. Soc.* 101:447–504.

McCafferty, W. P., and G. F. Edmunds, Jr. 1973. Subgeneric classification of *Ephemera* (Ephemeroptera: Ephemeridae). *Pan-Pac. Entomol.* 49:300–7.

McCafferty, W. P., and A. V. Provonsha. 1975. Reinstatement and biosystematics of *Heterocloeon* McDunnough (Ephemeroptera: Baetidae). *J. Georgia Entomol. Soc.* 10:123–27.

McDunnough. J. 1921. Two new Canadian mayflies (Ephemeridae). *Can. Entomol.* 53:117–20.

_____. 1925. New Canadian Ephemeridae with notes. *Can. Entomol.* 57:168–76.

_____. 1936a. A new arctic Baetid (Ephemeroptera). *Can. Entomol.* 68:32–34.

_____. 1936b. Further notes on the genus *Ameletus* with descriptions of new species (Ephemeroptera). *Can. Entomol.* 68:207–11.

_____. 1938. New species of North American Ephemeroptera with critical notes. *Can. Entomol.* 70:23–34.

_____. 1939. New British Columbian Ephemeroptera. *Can. Entomol.* 71:49–54.

_____. 1942. An apparently new *Thraulodes* from Arizona (Ephemerida). *Can. Entomol.* 74:117.

_____. 1943. A new *Cinygmula* from British Columbia (Ephemeroptera). *Can. Entomol.* 75:3.

Mayo, V. K. 1939. New western Ephemeroptera. *Pan-Pac. Entomol.* 15:145–54.

_____. 1951. New western Ephemeroptera. II. *Pan-Pac. Entomol.* 27:121–25.

_____. 1952a. New western Ephemeroptera. III. *Pan-Pac. Entomol.* 28:93–103.

_____. 1952b. New western Ephemeroptera. IV. *Pan-Pac. Entomol.* 28:179–86.

_____. 1969. Nymphs of *Thraulodes speciosus* Traver with notes on a symbiotic chironomid (Ephemeroptera: Leptophlebiidae). *Pan-Pac. Entomol.* 45:103–12.

_____. 1972. New species of the genus *Baetodes* (Ephemeroptera: Baetidae). *Pan-Pac. Entomol.* 48:226–41.

_____. 1973. Four new species of *Baetodes* (Ephemeroptera: Baetidae). *Pan-Pac. Entomol.* 49:308–14.

Mills, H. B., W. C. Starrett, and F. C. Bellrose. 1966. Man's effect on the fish and wildlife of the Illinois River. *Ill. Natur. Hist. Surv. Biol. Notes*, No. 57.

Müller-Liebenau, I. 1970. Revision der europäischen Arten der Gattung *Baetis* Leach, 1815 (Insecta, Ephemeroptera). *Gewässer und Abwässer* 48/49:1–214.

———. 1973. Morphological characters used in revising the European species of the genus *Baetis* Leach. In *Proceedings of the First International Conference on Ephemeroptera*, W. L. Peters and J. G. Peters, eds., pp. 182–98, E. J. Brill, Leiden.

———. 1974. *Rheobaetis*: A new genus from Georgia (Ephemeroptera: Baetidae). *Ann. Entomol. Soc. Amer.* 67:555–67.

Murphy, H. 1922. Notes on the biology of some of our North American species of mayflies. *Bull. Lloyd Lib.* 22, Entomol. Ser., 2:1–46.

Navas, P. L. 1924. Insectos de la America Central. *Broteria (Ser. Zool.)* 21:55–86.

———. 1935. Decadas de insectos nuevos 27. *Broteria (Ser. Zool.)* 31:97–107.

Neave, F. 1932. A study of the May flies (*Hexagenia*) of Lake Winnipeg. *Contr. Can. Biol. Fish.* 15 (Ser. A), 12:179–201.

Needham, J. G., and H. E. Murphy. 1924. Neotropical mayflies. *Bull. Lloyd Lib.* 24, Entomol. Ser., 4:1–79.

Needham, J. G., and P. R. Needham. 1962. *A guide to the study of freshwater biology*, 5th ed. Holden-Day, San Francisco.

Needham, J. G., J. R. Traver, and Y. C. Hsu. 1935. *The biology of mayflies with a systematic account of North American species*. Comstock, Ithaca.

Packer, J. S. 1966. A preliminary study of the mayflies of Honduras. *Ceiba* 12:1–10.

Pescador, M. L. 1973. The ecology and life history of *Baetisca rogersi* Berner (Ephemeroptera: Baetiscidae). In *Proceedings of the First International Conference on Ephemeroptera*, W. L. Peters and J. G. Peters, eds., pp. 211–15. E. J. Brill, Leiden.

Pescador, M. L., and W. L. Peters, 1971. The imago of *Baetisca becki* Schneider and Berner (Baetiscidae: Ephemeroptera). *Fla. Entomol.* 54:329–34.

———. 1974. The life history and ecology of *Baetisca rogersi* Berner (Ephemeroptera: Baetiscidae). *Bull. Fla. State Mus., Biol. Sci. Ser.*, 17:151–209.

Peters, W. L. 1959. A new species of *Callibaetis* from Kansas (Ephemeroptera: Baetidae). *J. Kans. Entomol. Soc.* 32:173–75.

———. 1969. *Askola froehlichi*, a new genus and species from southern Brazil (Leptophlebiidae: Ephemeroptera). *Fla. Entomol.* 52:253–58.

———. 1971. A revision of the Leptophlebiidae of the West Indies (Ephemeroptera). *Smithsonian Contrib. Zool.* 62:1–48.

Peters, W. L., and G. F. Edmunds, Jr. 1961. The mayflies (Ephemeroptera) of the Navajo Reservoir Basin, New Mexico and Colorado. *Univ. Utah Anthropol. Pap.* 55:107–11.

Pictet, F. J. 1843. *Historie naturelle generale et particulière des insectes névroptères. Famille des Ephémérides*. Geneva and Paris.

Richardson, E. 1921. Changes in the bottom fauna of the middle Illinois River and its connecting lakes since 1913–1915 as a result of the increase, southward, of sewage pollution. *Ill. Natur. Hist. Surv. Bull.* 14(4):33–75.

Roback, S. S. 1966. The Catherwood Foundation Peruvian-Amazon Expedition. VI. Ephemeroptera nymphs. *Monogr. Acad. Natur. Sci. Philadelphia* 14:129–99.

Schneider, R. F. 1966. Mayfly nymphs from northwestern Florida. *Quart. J. Fla. Acad. Sci.* 29:202–06.

Schneider, R. F., and L. Berner. 1963. A new southeastern species of *Baetisca* (Ephemeroptera: Baetiscidae). *Fla. Entomol.* 46:183–87.

Scott, D. C., L. Berner, and A. Hirsch. 1959. The nymph of the mayfly genus *Tortopus* (Ephemeroptera: Polymitarcyidae). *Ann. Entomol. Soc. Amer.* 52:205–13.

Spieth, H. T. 1937. An oligoneurid from North America. *J. New York Entomol. Soc.* 45:139–45.

———. 1938a. A method of rearing *Hexagenia* nymphs (Ephemerida). *Entomol. News* 49:29–32.

———. 1938b. Two interesting mayfly nymphs with a description of a new species. *Amer. Mus. Novitates* 970:1–7.

————. 1938c. Taxonomic studies on Ephemerida. I. Description of new North American species. *Amer. Mus. Novitates* 1002:1–11.

————. 1940a. The North American ephemeropteran species of Francis Walker. *Ann. Entomol. Soc. Amer.* 33:324–38.

————. 1940b. Studies on the biology of the Ephemeroptera. II. The nuptial flight. *J. New York Entomol. Soc.* 48:379–90.

————. 1941a. The North American ephemeropteran types of the Rev. A. E. Eaton. *Ann. Entomol. Soc. Amer.* 34:87–98.

————. 1941b. Taxonomic studies on the Ephemeroptera. II. The genus *Hexagenia*. *Amer. Midland Natur.* 26:233–80.

————. 1943. Taxonomic studies on the Ephemeroptera. III. Some interesting ephemerids from Surinam and other Neotropical localities. *Amer. Mus. Novitates* 1244:1–13.

————. 1947. Taxonomic studies on the Ephemeroptera. IV. The genus *Stenonema*. *Ann. Entomol. Soc. Amer.* 40:87–122.

Thew, T. B. 1956. A list of the mayflies of Iowa. *Museum Quarterly* (Davenport Public Museum) 1:1–6.

————. 1957. Observations on the oviposition of a species of Baetis. *Museum Quarterly* (Davenport Public Museum) 2:6.

————. 1958. Dodd's types of two species of *Callibaetis* (Ephemeroptera). *Entomol. News* 69:137.

————. 1959. Reexamination of some Nearctic species of the genus *Callibaetis* Eaton, with the description of a new species (Ephemeroptera: Baetidae). *Trans. Amer. Entomol. Soc.* 84:261–72.

Tiensuu, L. 1935. On the Ephemeroptera-fauna of Laatokan Karjala (Karelia Ladogensis). *Ann. Entomol. Fenn.* 1:3–23.

Traver, J. R. 1937. Notes on mayflies of the southeastern states (Ephemeroptera). *J. Elisha Mitchell Sci. Soc.* 53:27–86.

————. 1938. Mayflies of Puerto Rico. *J. Agr. Univ. Puerto Rico* 22:5–42.

————. 1940. *Compendium of entomological methods. I. Collecting mayflies (Ephemeroptera)*. Wards Natural Science Establishment, Rochester, N.Y.

————. 1944. Notes on Brazilian mayflies. *Bol. Mus. Nac. Brasil Zool.* 22:1–52.

————. 1946. Notes on Neotropical mayflies. I. Family Baetidae, subfamily Leptophlebiinae. *Rev. de Entomol.* 17:418–36.

————. 1947a. Notes on Neotropical mayflies. II. Family Baetidae, subfamily Leptophlebiinae. *Rev. de Entomol.* 18:149–60.

————. 1947b. Notes on Neotropical mayflies. III. Family Ephemeridae. *Rev. de Entomol.* 18:370–95.

————. 1950. Notes on Neotropical mayflies. IV. Family Ephemeridae (continued). *Rev. de Entomol.* 21:593–614.

————. 1956a. The genus *Asthenopodes* (Ephemeroptera). *Commun. Zool. Mus. Hist. Natur. Montevideo* 4:1–10.

————. 1956b. A new genus of Neotropical mayflies (Ephemeroptera: Leptophlebiidae). *Proc. Entomol. Soc. Wash.* 58:1–13.

————. 1958a. The subfamily Leptohyphinae (Ephemeroptera: Tricorythidae). *Ann. Entomol. Soc. Amer.* 51:491–503.

————. 1958b. Some Mexican and Costa Rican mayflies. *Bull. Brooklyn Entomol. Soc.* 53:81–89.

————. 1959a. Uruguayan mayflies. Family Leptophlebiidae: I. *Rev. Soc. Uruguaya Entomol.* 3:1–13.

————. 1959b. The subfamily Leptohyphinae. II. Five new species of *Tricorythodes* (Ephemeroptera: Tricorythidae). *Proc. Entomol. Soc. Wash.* 61:121–31.

————. 1960a. Some Mexican and Costa Rican mayflies. *Bull. Brooklyn Entomol. Soc.* 55:16–23.

────. 1960b. Uruguayan mayflies, Family Leptophlebiidae. III. *Rev. Soc. Uruguaya Entomol.* 4:73–86.

────. 1962. *Cloeon dipterum* (L.) in Ohio (Ephemeroptera: Baetidae). *Bull. Brooklyn Entomol. Soc.* 57:47–50.

────. 1964. A new species of the subgenus *Iron* from Mexico (Ephemeroptera: Heptageniidae). *Bull. Brooklyn Entomol. Soc.* 59/60:23–29.

Traver, J. R., and G. F. Edmunds, Jr. 1967. A revision of the genus *Thraulodes* (Ephemeroptera: Leptophlebiidae) *Misc. Publ. Entomol. Soc. Amer.* 5:349–95.

────. 1968. A revision of the Baetidae with spatulate-clawed nymphs (Ephemeroptera). *Pac. Insects* 10:629–77.

Trost, L. M. W., and L. Berner. 1963. The biology of *Callibaetis floridanus* Banks (Ephemeroptera: Baetidae). *Fla. Entomol.* 46:285–99.

Tshernova, O. A. 1948. A new genus and species of mayfly from the Amur Basin (Ephemeroptera, Ametropodidae). *Dokl. Akad. Nauk SSSR* 60:1453–55.

Ulmer, G. 1920a. Über die Nymphen einiger exotischen Ephemeropteren. *Festschrift für Zschokke.* 25:3–25.

────. 1920b. Neue Ephemeropteren. *Archiv. für Naturgeschichte* 85 (Abteilung A11):1–80.

────. 1942. Alte und neue Eintagsfliegen (Ephemeropteren) aus Sud- und Mittelamerika. I. *Stett. Entomol. Zeit.* 103:98–128.

────. 1943. Alte und neue Eintagsfliegen (Ephemeropteren) aus Sud- und Mittelamerika. II. *Stett. Entomol. Zeit.* 104:14–46.

Upholt, W. M. 1936. A new species of mayfly from California (Ephemerida: Baetidae). *Pan-Pac. Entomol.* 12:120–22.

────. 1937. Two new mayflies from the Pacific Coast. *Pan-Pac. Entomol.* 13:85–88.

Walker, F. 1853. *List of neuropterous insects in the British Museum. III. Termites and Ephemeridae.* British Museum (Natural History), London.

Wood, K. G. 1973. Decline of *Hexagenia* (Ephemeroptera) nymphs in Western Lake Erie. In *Proceedings of the First International Conference on Ephemeroptera*, W. L. Peters and J. G. Peters, eds., pp. 26–32. E. J. Brill, Leiden.

Wright, M., and L. Berner. 1949. Notes on mayflies of eastern Tennessee. *J. Tenn. Acad. Sci.* 24:287–98.

INDEX

Index

Page numbers in italic type refer to illustrations; those in boldface type identify the principal references to taxonomic groups. Species listed in tables are not included in the index.

325